Collins

AQA GCSE Revision

English

Poetry Anthology: Power & Conflict

English

D0256167

AQA GCSE

Poetry Revision Guide

Rachel Grant

Anthology ► Contents

Contents

	Revise	Practise	Review

Overview of Poetry Anthology Cluster: Power and Conflict

- The poems in this cluster focus on the various aspects of power and conflict.
- Some poems, such as *The Charge of the Light Brigade* and *Bayonet Charge*, focus on warfare and battlefields, while others, such as *Storm on the Island*, consider conflict in nature.
- Some poems discuss how conflict can change the perspectives of individuals for ever. In *Remains*, Armitage considers the after-effects of conflict on a combatant.
- Other poems, such as *My Last Duchess*, consider the use and effects of power in marriage, while others, such as *The Emigrée* and *Checking Out Me History*, look at how identity is shaped by **culture**, history and political oppression.

 Key Point

The poems in the Power and Conflict cluster focus on aspects and themes of power and conflict. Many are about wars and their effects; others are about the power of nature or political power.

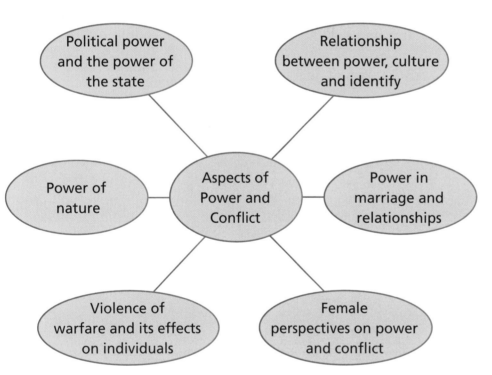

Political power and the power of the state

Relationship between power, culture and identify

Power of nature

Aspects of Power and Conflict

Power in marriage and relationships

Violence of warfare and its effects on individuals

Female perspectives on power and conflict

Different Viewpoints

- Some poems in the Power and Conflict cluster are about an incident in war, or armed conflict, re-told by someone involved in it. *Exposure*, for example, is based on the poet's experience in the First World War.
- Some poets have written about events in which they were not personally involved.
    - For example, Tennyson wrote *The Charge of the Light Brigade* after reading about the charge in a newspaper.
    - Other poems, such as *Poppies*, consider the effects of warfare on those not directly involved in conflict.
- *Bayonet Charge* was inspired by an incident in the First World War, but was written long afterwards and the soldier's experience is narrated in the third-person. In *Remains* the speaker is a soldier suffering from post-traumatic stress disorder (PTSD) after serving in the Iraq war.
- In *Kamikaze*, a daughter speaks of her father's experience in the Second World War and how it shaped his whole life afterwards.

Themes

- Some poems do not refer to particular conflicts, but offer general comment on the **theme**.
    - *War Photographer* is about attitudes to war in general and in particular the way images of war are used by the media.
    - Weir's *Poppies* gives a mother's perspective on the loss of a son who is a soldier.
- Several poems explore ideas about the power of individuals and how they use that power.
- Browning's *My Last Duchess* provides a chilling example of the abuse of absolute power in Renaissance Italy.
- Shelley's *Ozymandias* considers how the power of time and nature will always triumph over human power and dominion, no matter how invincible they may seem.
- In *Checking Out Me History*, Agard talks about how he intends to take back power and re-gain control over his own identity.
- In *Tissue* and *Checking Out Me History*, the poets explore the relationships between identity, power and culture.
- Political messages are closely identified with power. Blake's *London* is a protest against the misery and disempowerment of people that he sees all around him.
- Wordsworth's *Extract from The Prelude*, and *Storm on the Island* by Heaney, consider nature's awe-inspiring power in very different contexts.

Key Words
Culture
Theme

Cluster Overview: Power and Conflict

Effects of Voice, Mood, Technique and Form

- The poems in the Power and Conflict cluster encompass a great range of voices, moods, techniques and forms.
- Some of the poems are angry or sorrowful, others resigned and bitter, and still others show sympathy for the oppressed and hope for the future.
- You will find revulsion at the violence of war and the misuse of power, as well as admiration for those who show bravery and resilience.
- Pay close attention to the structure, language and form the poets use to add effect and meaning.

Utilising the Various Poems, Ideas and Themes

- In the exam you will need to make connections between poems, looking at similarities and differences in the content, and in how poets present their ideas about power and conflict.
- The table opposite is a short guide to the themes presented in the poems. It may be used as a starting point, but it is not exhaustive.
- All the poems in the cluster are multi-layered and are open to many interpretations that highlight different aspects of several themes. They should be read and considered as such.

 Key Point

All the poems in the *Power and Conflict* cluster can be interpreted in different ways.

 Key Words

Voice
Mood
Form

Poem	Aspect(s) of Theme Presented	Additional Comments including Method and Form
Ozymandias	– How power declines	– Sonnet
London	– Powerlessness	– Strong and regular rhythm and rhyme – Ballad form
Extract from The Prelude	– The power of the natural world	– Iambic pentameter – Personification
My Last Duchess	– Power relations in a marriage	– Dramatic monologue – Rhyming couplets with considerable use of enjambment
The Charge of the Light Brigade	– Patriotism – Courage	– Strong use of rhythm and repetition – Conclusive final verse
Exposure	– Futility of war – Nihilism	– Use of half-rhyme – Repetition
Storm on the Island	– The power of the natural world	– Extended metaphor of a military attack
Bayonet Charge	– A soldier's perspective on conflict	– Second verse suggests a pause in the action
Remains	– Psychological effects of conflict on soldiers	– Two halves represent the soldier's experience
Poppies	– A mother's perspective on conflict	– Dramatic monologue – Sentences starting in the middle of a line
War Photographer	– How conflict is presented in the media	– Rhyming couplets in each verse
Tissue	– The relationships between fragility and power	– Free verse; enjambment across verse boundaries. – Striking final image set apart
The Emigrée	– Individual freedom versus forced absence/isolation	– Longer sentences contrast with shorter sentences – Extended metaphor
Kamikaze	– A female perspective on conflict	– Italics mark direct speech of personal memory compared to what is speculated
Checking Out Me History	– Relationship between culture, power and identity	– Strong use of repetition/refrain and rhythm – Two types of verse presentation

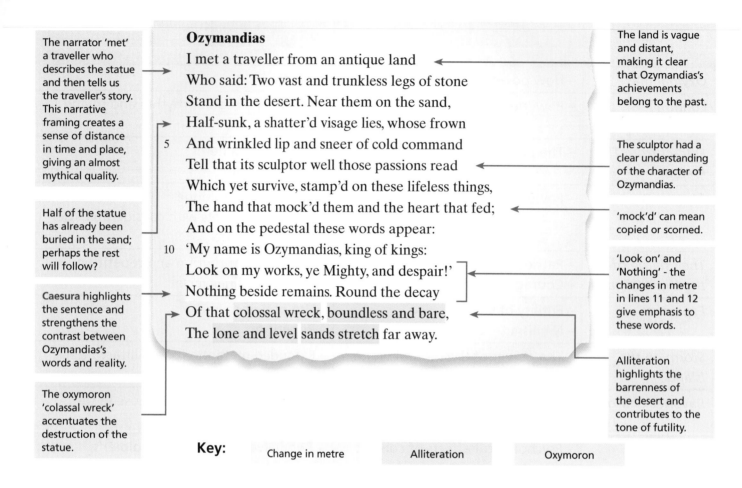

The narrator 'met' a traveller who describes the statue and then tells us the traveller's story. This narrative framing creates a sense of distance in time and place, giving an almost mythical quality.

Half of the statue has already been buried in the sand; perhaps the rest will follow?

Caesura highlights the sentence and strengthens the contrast between Ozymandias's words and reality.

The oxymoron 'colassal wreck' accentuates the destruction of the statue.

Ozymandias

I met a traveller from an antique land
Who said: Two vast and trunkless legs of stone
Stand in the desert. Near them on the sand,
Half-sunk, a shatter'd visage lies, whose frown
5 And wrinkled lip and sneer of cold command
Tell that its sculptor well those passions read
Which yet survive, stamp'd on these lifeless things,
The hand that mock'd them and the heart that fed;
And on the pedestal these words appear:
10 'My name is Ozymandias, king of kings:
Look on my works, ye Mighty, and despair!'
Nothing beside remains. Round the decay
Of that colossal wreck, boundless and bare,
The lone and level sands stretch far away.

The land is vague and distant, making it clear that Ozymandias's achievements belong to the past.

The sculptor had a clear understanding of the character of Ozymandias.

'mock'd' can mean copied or scorned.

'Look on' and 'Nothing' - the changes in metre in lines 11 and 12 give emphasis to these words.

Alliteration highlights the barrenness of the desert and contributes to the tone of futility.

Key: Change in metre   Alliteration   Oxymoron

About the Poem

- The poem was written by Percy Bysshe Shelley in 1818. It is thought that the poem was inspired by the statue of the powerful Egyptian ruler Rameses II (Ozymandias in the poem), which was brought to London about that time.
- The narrator recounts an anonymous traveller's tale.
- The narrator re-tells the traveller's description of the remains of a vast, ruined statue of a powerful ruler, half-buried in the desert.
- The statue is a **metaphor** which reveals the character and foolishness of the ruler.
- There are two voices in the poem – the traveller and Ozymandias. We learn about two characters: the artist and Ozymandias. Ozymandias's words are in inscription on the pedestal.
- The poem can be seen as an analogy: time will challenge any great civilisation or ruler.

Ideas, Themes and Issues

- **Death and mortality:** The desert symbolises the passing of time. It erases all traces of Ozymandias and reminds us that death comes to us all.
- **Power:** Ozymandias's power was absolute. The face of the statue and the words on the pedestal reveal his character. We feel no sympathy for him.

- **Pride:** The vast size of the statue reflects the ruler's pride. The futility of his pride is demonstrated by the ruin of the statue and his achievements. His name is significant as it is derived from the Greek 'ozium' (air) and 'mandate' (rule), meaning he is 'ruler of nothing'.
- **Art:** The achievements of the sculptor outlive those of his subject.

Form, Structure and Language

- The poem is a **sonnet** written in **iambic pentameter**. Traditionally sonnets are love poems, but here political ideas are explored. The metre changes in lines 11 and 13, which helps to emphasise the futility of Ozymandias's boast.
- Like many sonnets, this poem can be divided into three parts:
  - The traveller creates a picture of Ozymandias in the first and final parts.
  - The middle part celebrates the sculptor's artistry and explores the idea of art as enduring. The irony is that the sculptor's art endures, while Ozymandias's power has not – it has declined, but the sculptor's goes on.
- The use of the **oxymoron** in line 13 emphasises the destruction of the statue.
- When Ozymandias claims the title 'king of kings' (10) it reveals his pride and contrasts ironically with the literal meaning of his name.

Quick Test

1. How do the setting and narrator contribute to our sense of Ozymandias's diminished power?
2. What does the poem suggest about the nature of power?
3. How does the poet create a negative impression of the character of Ozymandias?

Key Words

Caesura
Metaphor
Sonnet
Iambic pentameter
Oxymoron

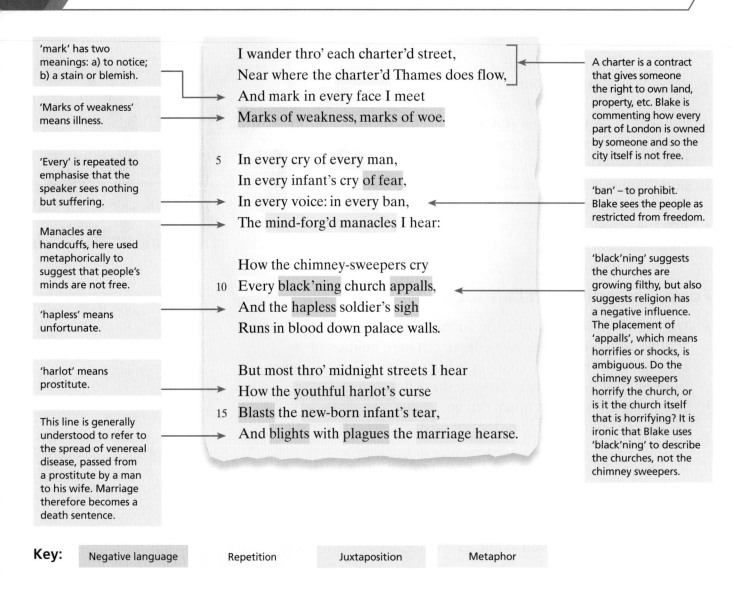

'mark' has two meanings: a) to notice; b) a stain or blemish.

'Marks of weakness' means illness.

'Every' is repeated to emphasise that the speaker sees nothing but suffering.

Manacles are handcuffs, here used metaphorically to suggest that people's minds are not free.

'hapless' means unfortunate.

'harlot' means prostitute.

This line is generally understood to refer to the spread of venereal disease, passed from a prostitute by a man to his wife. Marriage therefore becomes a death sentence.

A charter is a contract that gives someone the right to own land, property, etc. Blake is commenting how every part of London is owned by someone and so the city itself is not free.

'ban' – to prohibit. Blake sees the people as restricted from freedom.

'black'ning' suggests the churches are growing filthy, but also suggests religion has a negative influence. The placement of 'appalls', which means horrifies or shocks, is ambiguous. Do the chimney sweepers horrify the church, or is it the church itself that is horrifying? It is ironic that Blake uses 'black'ning' to describe the churches, not the chimney sweepers.

I wander thro' each charter'd street,
Near where the charter'd Thames does flow,
And mark in every face I meet
Marks of weakness, marks of woe.

5 In every cry of every man,
In every infant's cry of fear,
In every voice: in every ban,
The mind-forg'd manacles I hear:

How the chimney-sweepers cry
10 Every black'ning church appalls,
And the hapless soldier's sigh
Runs in blood down palace walls.

But most thro' midnight streets I hear
How the youthful harlot's curse
15 Blasts the new-born infant's tear,
And blights with plagues the marriage hearse.

Key: Negative language   Repetition   Juxtaposition   Metaphor

About the Poem

- The poem was written by William Blake and published in his collection *Songs of Experience* in 1794.
- The poem reflects Blake's views about the effect of the Industrial Revolution on England.
- Written in the **first person**, the poem details a walk through London. Blake is angry that everyone he sees is suffering and that no one is doing anything about it. He blames the church and the State for the situation. In the poem, Blake protests at the lack of individual freedom he sees everywhere around him.

Ideas, Themes and Issues

- **Social comment:** Blake sees London as a place full of despair and misery because of the oppression people are under. He does not see any hope for society because people have so little power that they lack the imagination to change their lives. Every strata of society is mentioned in the poem: chimney-sweepers (9), harlots (14), new-born babies (15), and institutions such as the church (10), army (11), royalty (12) and marriage (16).
- **Despair:** The church and the state offer no hope for the poor and, in fact, are the instruments of its oppression. Blake sees no hope for the future either, since children born into such conditions are blighted by birth to remain there and marriage is also blighted 'with plagues'(16).

Form, Structure and Language

- Blake uses negative language throughout to show his view of Londoners' plight. **Juxtaposition** of the ideas of innocence and experience shocks the reader.
- **Repetition** ('every', 'In every') is used relentlessly to drive home the bleakness of Blake's message and he uses **emotive** language to make the reader feel as angry and sorrowful as he himself feels.
- Blake uses a simple, repetitive, rhyming structure of a ballad form. The ballad was a popular form because it is memorable and helped people to share stories and express ideas.
- The **metaphor** used in line 8 ('mind-forg'd manacles) suggests that people are trapped as much by their own attitudes as by society. Convicts wore manacles, so this is an image of powerlessness. The word 'forg'd' can also mean fake, suggesting that the things that trap people are not real, if only they could see it.
- Juxtaposition shows that every hope of happiness is tainted with despair. The repetitive metre reflects the idea that people's lives are monotonous.

Quick Test

1. What does Blake think are the main problems in London?
2. How does Blake show these ideas?
3. What do you think Blake's intention was in writing this poem?

Key Words

First person
Juxtaposition
Repetition
Emotive
Metaphor

'her' is nature. Romantic poets often personified nature.

'her' in this context refers to the boat: sea vessels are traditionally given female names, so the feminine pronouns are used.

He knew it was wrong to take the boat, hence 'troubled' (guilty) pleasure.

A 'pinnace' is a small, light boat; 'elfin' means small and delicate, like elves!; 'lustily' means vigorous.

'huge' is repeated for emphasis and contrasts with the 'elfin' boat and the young poet at the oars.

In contrast to his earlier vigorous, purposeful rowing, now his arms are trembling with nervous fear.

'stole' has two meanings: a) moved in a secretive, unobtrusive way; b) took without permission.

Forces of nature that are beyond understanding.

The image of the towering peak – and the impression it gave him of nature's power and awesomeness – continued to haunt him, day and night.

One summer evening (led by her) I found
A little boat tied to a willow tree
Within a rocky cove, its usual home.
Straight I unloosed her chain, and stepping in
5 Pushed from the shore. It was an act of stealth
And troubled pleasure, nor without the voice
Of mountain-echoes did my boat move on;
Leaving behind her still, on either side,
Small circles glittering idly in the moon,
10 Until they melted all into one track
Of sparkling light. But now, like one who rows,
Proud of his skill, to reach a chosen point
With an unswerving line, I fixed my view
Upon the summit of a craggy ridge,
15 The horizon's utmost boundary; far above
Was nothing but the stars and the grey sky.
She was an elfin pinnace; lustily
I dipped my oars into the silent lake,
And, as I rose upon the stroke, my boat
20 Went heaving through the water like a swan;
When, from behind that craggy steep till then
The horizon's bound, a huge peak, black and huge,
As if with voluntary power instinct,
Upreared its head. I struck and struck again,
25 And growing still in stature the grim shape
Towered up between me and the stars, and still,
For so it seemed, with purpose of its own
And measured motion like a living thing,
Strode after me. With trembling oars I turned,
30 And through the silent water stole my way
Back to the covert of the willow tree;
There in her mooring-place I left my bark, -
And through the meadows homeward went, in grave
And serious mood; but after I had seen
35 That spectacle, for many days, my brain
Worked with a dim and undetermined sense
Of unknown modes of being; o'er my thoughts
There hung a darkness, call it solitude
Or blank desertion. No familiar shapes
40 Remained, no pleasant images of trees,
Of sea or sky, no colours of green fields;
But huge and mighty forms, that do not live
Like living men, moved slowly through the mind
By day, and were a trouble to my dreams.

'Straight' in this context means 'straight away' or immediately. He didn't think about it: he just did it.

'stealth' means done in secret.

Beautiful description of the ripples that the boat leaves in the still lake and the moon's reflection in it. The assonance of 'i' and alliteration of 't' and 's' convey the still and quiet beauty of the scene.

At first, his rowing is energetic and determined.

'instinct' is an inborn behaviour; here Wordsworth is giving the peak an innate power as if it is a living organism.

The peak seems to grow as the boat gets nearer to it.

'covert' has two meanings: a) thicket or copse; b) secret.

'bark' is another word for boat.

The effect of what had happened, with the peak seeming to warn him for stealing the boat, preys on his mind.

He felt depressed!

Key:

Simile

Personification

Repetition

About the Poem

- The poem was written by William Wordsworth. It is an extract from his long autobiographical poem *The Prelude*, which he started to write in 1798, when he was 28.
- Wordsworth called it 'a poem on the growth of my own mind'. The final version consists of 14 books.
- This extract describes how as a young boy, Wordsworth stole a boat and was taught a lesson by nature. The experience gave him a lasting impression of nature's power and mystery.

Ideas, Themes and Issues

- **The power and beauty of nature:** Nature is presented initially as benign and beautiful as the boy enjoys his joyride in the boat. He enjoys the view of the lake and the stars, the ripples from the boat and the moon. But then, as he rows with his eyes fixed on one crag, another crag appears behind it. Here the mood changes and nature becomes frightening and awe-inspiring: 'a huge peak, black and huge' (22) towers over him and the harder he rows, the bigger it becomes – it seems to grow in front of his eyes. Even as a boy Wordsworth knew, rationally, that the size of the peak only appears to grow ('As if' (23) and 'For so it seemed' (27) tell us this), but his tranquil mood has been broken. Subdued by nature's immensity, he turns around 'With trembling oars' (29) and returns the boat. The different perspectives of the **imagery** used reflects this change: from 'Small circles' (9) on the lake to the immensity of the black peak, 'the grim shape/Towered up' (25–26).
- **The power of memory and imagination:** The after-effects of his boyhood experience give purpose to the older poet's recollection of the incident. Immediately after returning to the boat he felt 'grave/And serious' (33–34), but afterwards he was unable to shake off the feeling and the memory of the peak, and the feelings of awe that it inspired in him. It left a lasting impression and shaped his view of nature from then on. The older poet views and recalls this incident as a mind-forming moment in his life, because it was when he first became aware 'Of unknown modes of being' (37).

Form, Structure and Language

- The poet uses **similes** that compare natural objects to living things and vice versa (the peak is 'like a living thing' (28) and the boat is 'like a swan' (20)).
- **Personification** and imagery is used to convey the beauty, but also the dangerous power and awesomeness, of nature.
- **Conjunctions** (*and*, *but*, *or*, *for*, *since*) link the events together seamlessly and use of **enjambment** encourage us to keep reading through the sequence of recalled events.
- The poem is written in **iambic pentameter**, a verse form that is very close in rhythm to the speaking voice and which is therefore suitable for this very personal poem.

 Key Point

The notion that there are forces at work in nature that we can never understand is an important idea in Romantic poetry.

 Key Words

Imagery
Similes
Personification
Conjunctions
Enjambment
Iambic pentameter

 Quick Test

1. Where does the narrative begin?
2. What time of day is it?
3. What did the speaker do when he found the boat?

About the Poem

- The poem was written by Robert Browning and was first published in 1842.
- In the poem, we hear the Duke describing a portrait of his late wife to a Count's envoy, who is helping to arrange a marriage between the Duke and his master's daughter.
- The Duke unwittingly reveals the true character of both himself and the Duchess through the things he says about her.
- We learn that he had his wife killed, as he suspected she was unfaithful to him. The poem explores power relations in a marriage.

Ideas, Themes and Issues

- **Violence, power and control:** The Duke's lack of remorse at having his wife murdered is shocking. He has absolute power, as is shown by the standards he refers to that she did not meet and that his pride would not allow him to talk to her about. No one dares to ask him anything about the portrait, so he has to make up what he thinks their questions would be. He is so obsessed with controlling his wife that he uses his power to have her killed.
- **Jealousy and pride:** The Duke was jealous of the attention his wife gave to others and wanted to be the only person in her favour. His own jealousy led him to suspect her of infidelity. The Duke boasts about the beautiful

Ferrara

That's my last Duchess painted on the wall,
Looking as if she were alive. I call
That piece a wonder, now; Frà Pandolf's hands
Worked busily a day, and there she stands.
5 Will't please you sit and look at her? I said
"Frà Pandolf" by design, for never read
Strangers like you that pictured countenance,
The depth and passion of its earnest glance,
But to myself they turned (since none puts by
10 The curtain I have drawn for you, but I)
And seemed as they would ask me, if they durst,
How such a glance came there; so, not the first
Are you to turn and ask thus. Sir, 'twas not
Her husband's presence only, called that spot
15 Of joy into the Duchess' cheek; perhaps
Frà Pandolf chanced to say, "Her mantle laps
Over my lady's wrist too much," or "Paint
Must never hope to reproduce the faint
Half-flush that dies along her throat": such stuff
20 Was courtesy, she thought, and cause enough
For calling up that spot of joy. She had
A heart—how shall I say?— too soon made glad,
Too easily impressed; she liked whate'er
She looked on, and her looks went everywhere.
25 Sir, 'twas all one! My favour at her breast,
The dropping of the daylight in the West,
The bough of cherries some officious fool
Broke in the orchard for her, the white mule
She rode with round the terrace—all and each

Frà Pandolf is the name of the painter.

'countenance' means face.

The line gives a positive description of the Duchess. (compare with lines 13–34).

A positive image, framed in a negative light by his perceptions.

The self-interruption suggests spontaneity, but this is a rehearsed speech that the Duke has given many times.

Implies the Duchess was unfaithful, or at least looked at other men in a way the Duke did not like.

Key:

| Repetition | Control |
| Duchess's qualities | Self-interruption |

30 Would draw from her alike the approving speech,
    Or blush, at least. She thanked men—good! but thanked
    Somehow—I know not how—as if she ranked
    My gift of a nine-hundred-years-old name
    With anybody's gift. Who'd stoop to blame
35 This sort of trifling? Even had you skill
    In speech—(which I have not)—to make your will
    Quite clear to such an one, and say, "Just this
    Or that in you disgusts me; here you miss,
    Or there exceed the mark"—and if she let
40 Herself be lessoned so, nor plainly set
    Her wits to yours, forsooth, and made excuse,
    E'en then would be some stooping; and I choose
    Never to stoop. Oh, sir, she smiled, no doubt,
    Whene'er I passed her; but who passed without
45 Much the same smile? This grew; I gave commands;
    Then all smiles stopped together. There she stands
    As if alive. Will't please you rise? We'll meet
    The company below, then. I repeat,
    The Count your master's known munificence
50 Is ample warrant that no just pretense
    Of mine for dowry will be disallowed;
    Though his fair daughter's self, as I avowed
    At starting, is my object. Nay, we'll go
    Together down, sir. Notice Neptune, though,
55 Taming a sea-horse, thought a rarity,
    Which Claus of Innsbruck cast in bronze for me!

Annotations:

- She was gracious and grateful, but the Duke sees this as a flaw.
- Reveals what the Duke values – status.
- Suggests a set of standards (his) which must be met.
- Implies that for him to correct her behaviour would be demeaning to him.
- Caesura highlights the sinister euphemism.
- His future wife is also an 'object', which is chilling considering the content of his speech.
- Claus is the name of the sculptor of the bronze sea-horse.

objects he owns, including the Duchess's portrait, unaware of the effect of his words. He is proud of his name and felt the Duchess didn't respect its value.

Form, Structure and Language

- The poem is a **dramatic monologue**.
- The **first person** narrative allows us to fully understand the Duke's actions and motivations, perhaps even when he doesn't recognise them himself.
- The descriptions of the Duchess reveal a kind and gracious woman, which makes us sympathetic towards her and shows the depths of the Duke's need to control her.
- Repetition suggests the Duke's preoccupation with certain ideas and behaviours. The Duke's narcissism and pride is suggested by the use of **possessive pronouns** (mine, your).
- The poem is written in **rhyming couplets**, but because the lines are not **end-stopped** the rhyme is less obvious than it might be. This pushes the poem on, as the Duke relentlessly pursues his next bride.

Quick Test

1. How does the structure of the poem reflect the Duke's present concerns?
2. Why do we learn about the Duke's actions from his own mouth?
3. Why does the poem end with the Duke describing another piece of art?

Key Words

Dramatic monologue
First person
Possessive pronoun
Rhyming couplet
End-stopped

Ozymandias

1 Who is speaking for most of the poem?

The traveller of say w. [1]

2 What poetic form does Shelley use?

sonnet [1]

3 Which details tell us about Ozymandias's character?

'King of kings'   'the hand that fed' [1]

4 What is the literal meaning of Ozymandias's name? Why is this ironic?

[2]

5 Does the traveller have more admiration for the sculptor or his subject? Explain.

has more admiration for the sculptor as it has lasted through time whereas ozymandias - 'nothing beside remain' [2]

6 What do the words 'antique land' in line 1 suggest?

Old, an era that has passed, [1]

7 There is irony in the final line. Explain where you think the irony lies.

It lacks any characters, removes ozymandias. [1]

8 There are three voices in the poem. Who are they and what is one effect of having this many voices?

Shelly's voice 'I met', the travellers voice 'he said', Ozymandias 'I am'. [2]

9 In what way could Ozymandias be said to live on, even though his kingdom has disappeared?

his statue has lived though time. [1]

10 Who do you think Ozymandias refers to by 'ye Mighty' (11)?

because he is hubristic. [1]

London

① What picture of life in London is presented in the poem?

the picture of the poor + unjust life. [1]

② How would you describe the poet's mood? Choose two from the following:

(angry)   uplifted   amused   hopeful   (depressed) [1]

③ The word 'every' is repeated six times over five lines. What is one effect of the repetition of this word?

takes in the vast nature of the problem. [1]

④ The poem is tightly structured in terms of rhythm and rhyme. What is one effect of this?

reflects the tight structure of his 'walk' [1]

⑤ In the third verse, how does Blake encompass the whole of society in his depiction of London?

Use of church, use of poor, use of soldiers, use of palace/royalty, encapturing all of society. [1]

⑥ Why do you think Blake used the adjective 'black'ning' to describe the church?

The 'good' of the church is often associated with light. [2]

⑦ Pick out another interesting adjective that Blake uses and analyse its effect.

'blights' and 'plagues', describing marriage (religious) 'bl' plosive sound 2 adjectives overwhelm sentence like sickness [2]

⑧ Do you have a sense that Blake blames anyone or anything for the miserable state of London's population?

'runs in blood down palace walls' - 'Blac'ning church'. [1]

⑨ Blake focuses particularly on the plight of children. Why do you think he does this?

the children have a chance to change the future. [1]

⑩ What do you understand by the phrase 'mind-forged manacles' (8)?

[1]

Extract From The Prelude

1 What explanation does the boy give for finding the boat?

he was 'bled (by her)'

[1]

2 Find a phrase that indicates that the boy acted furtively, in secret.

'It was an act of stealth'

[1]

3 The boy took the boat without permission. Is this important, given what happens? Explain.

It shows the 'punishment', but also shows how he has chosen to stray into the water.

[2]

4 What is the boy's mood at the start of his trip?

apprehensive.

[1]

5 How does he navigate the boat?

using oars + current.

[1]

6 Pick out one image that suggests he was a strong oarsman.

'unswerving line'.

[1]

7 Which line marks a change in the tone of the poem?

'a huge peak, black and huge'

[1]

8 Why does the huge peak seem to grow in size?

he is oaring further and further away, causing him to notice a second peak.

[1]

9 How did the experience affect the boy afterwards?

more scared, mature?, acts less nonchalently.

[1]

10 What do you understand by the lines, ' But huge and mighty forms, that do not live/ Like living men,' (42–43)?

[2]

My Last Duchess

1 Who is the Duke speaking to in the poem? Why is he visiting the Duke?

A spokesperson for someone powerful. [2]

2 How do we know he has given this speech before?

'they' presence. [1]

3 What do we learn of the Duchess's character?

She was 'too eager' 'too soon made glad' [1]

4 Pick out an example from the Duke's monologue that shows his obsession with power and status.

ITS A MONOLOGUE?.? 'Frà pandolf' [1]

5 What is one effect of the Duke's use of caesura in lines 45-46?

[1]

6 Give two examples of the Duke's use of colloquial language. Why does Browning choose to use this?

It almost makes him feel less human and more psychopathic. [3]

7 What was it about the Duchess's behaviour that particularly enraged the Duke?

'too soon made glad' 'she would laugh with other people. [1]

8 What is the effect of the Duke's long description of the Duchess's appearance in the painting?

Shows his controlling nature - 'gave commands'. [1]

9 Pick out two phrases that suggest the Duchess is still alive for the Duke, through the painting.

'As if alive' [2]

10 What poetic form does Browning use here? Identify three features of this form.

Dramatic monologue - single speaker.
- single paragraph.
- description. [4]

About the Poem

- The poem was written by Alfred Lord Tennyson in 1854. Based on a real event in the Crimean War, the poem was inspired by a newspaper account.
- The poem opens at the start of the charge. There has been a mistake with the order to charge, but the patriotic cavalry ride on fearlessly, despite having fewer men and poorer weapons than the enemy.
- At the end, Tennyson calls for the cavalrymen to be remembered for their courage.

Ideas, Themes and Issues

- **Patriotism:** Tennyson celebrates the unthinking patriotism of the cavalry. They knew they were facing almost certain defeat and probable death, but they obeyed the order to charge.
- **Conflict:** The poem vividly evokes the violence of the battle which was waged with cannon and sabres. The cavalry faced enemy gunfire from all sides as they rode through the valley. At the end of the valley they turned and rode back again, meaning that those who survived withstood enemy fire twice. Tennyson emphasises the glory of the conflict, though it ended in defeat, and praises the courage, heroism and patriotism shown by the soldiers. He does not mention the fact that the deaths were pointless and gives only passing mention to the fact that the order to charge was a mistake.

Personification of the place and a Biblical reference ('the valley of the shadow of death' - Psalm 23).

The rhyme scheme varies throughout the poem.

Repetition emphasises the patriotism of the soldiers. They did not challenge the order.

Accumulation emphasises that they were outnumbered and were at a strategic disadvantage.

Verbs suggest the sounds of battle.

1.

Half a league, half a league,
Half a league onward,
All in the valley of Death
⌈   Rode the six hundred.
5 │ "Forward, the Light Brigade!
└ Charge for the guns!" he said.
Into the valley of Death
    Rode the six hundred.

2.

"Forward, the Light Brigade!"
10 Was there a man dismaye'd?
Not though the soldier knew
    Some one had blunder'd.
⌈ Theirs not to make reply,
│ Theirs not to reason why,
15 │ Theirs but to do and die.
└ Into the valley of Death
    Rode the six hundred.

The commander had made a mistake when he gave the order.

Repetition shows that the soldiers kept going despite the great danger.

3.

⌈ Cannon to right of them,
│ Cannon to left of them,
20 └ Cannon in front of them
⌈   Volley'd and thunder'd;
└ Storm'd at with shot and shell,
Boldly they rode and well,
Into the jaws of Death,
25 Into the mouth of Hell
    Rode the six hundred.

Key: Repetition   Descriptions of the valley   Descriptions of the battle

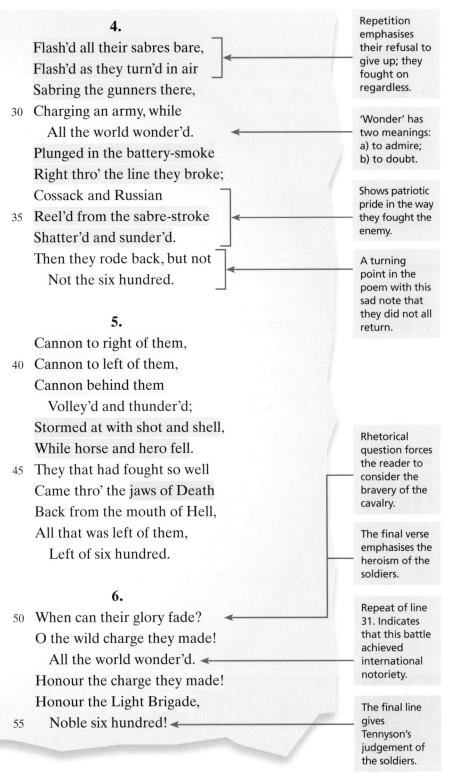

4.

Flash'd all their sabres bare,
Flash'd as they turn'd in air
Sabring the gunners there,
30 Charging an army, while
  All the world wonder'd.
Plunged in the battery-smoke
Right thro' the line they broke;
Cossack and Russian
35 Reel'd from the sabre-stroke
Shatter'd and sunder'd.
Then they rode back, but not
  Not the six hundred.

5.

Cannon to right of them,
40 Cannon to left of them,
Cannon behind them
  Volley'd and thunder'd;
Stormed at with shot and shell,
While horse and hero fell.
45 They that had fought so well
Came thro' the jaws of Death
Back from the mouth of Hell,
All that was left of them,
  Left of six hundred.

6.

50 When can their glory fade?
O the wild charge they made!
  All the world wonder'd.
Honour the charge they made!
Honour the Light Brigade,
55   Noble six hundred!

Repetition emphasises their refusal to give up; they fought on regardless.

'Wonder' has two meanings: a) to admire; b) to doubt.

Shows patriotic pride in the way they fought the enemy.

A turning point in the poem with this sad note that they did not all return.

Rhetorical question forces the reader to consider the bravery of the cavalry.

The final verse emphasises the heroism of the soldiers.

Repeat of line 31. Indicates that this battle achieved international notoriety.

The final line gives Tennyson's judgement of the soldiers.

Form, Structure and Language

- The poem is written in the **third person**. This has the effect of making the poem sound like an official account of the battle.
- The **rhythm** of the poem evokes the sound of hoof-beats. There are two main stressed syllables in each line at the beginning and in the middle of the line, followed by two unstressed syllables.
- **Repetition** is a strong feature, emphasising the relentless forward motion of the cavalry charge and how they unquestioningly follow the order.
- **Personification** emphasises the dangers the cavalry faced.
- Direct speech in the first and second verses reminds us that these were real soldiers in a real battle. The order given to the cavalry evokes a strong emotional response in the reader because it is as if we are hearing the order itself, first-hand.
- Tennyson's pride in the patriotism shown by the soldiers is evident in lines 23, 45 and in the sixth verse, in which the reputation of the cavalry is sealed for all time.

Quick Test

1. How does the poet suggest his admiration for the soldiers?
2. How does the poet suggest the reality of battle?
3. What does the final verse emphasise about the event?

Key Words

Third person
Rhythm
Repetition
Personification

Part I

The alliteration of the 's' sound and the assonance of 'i' makes us 'hear' the ice in the winds.

Our brains ache, in the merciless iced east winds that knife
    us ...

'knife' is used dramatically as a verb, meaning 'to stab'.

Wearied we keep awake because the night is silent ...
Low, drooping flares confuse our memory of the salient ...

5 Worried by silence, sentries whisper, curious, nervous,
        But nothing happens.

The 's' sound alliterated conjures the sound of whispering.

Watching, we hear the mad gusts tugging on the wire.
Like twitching agonies of men among its brambles.
Northward incessantly, the flickering gunnery rumbles,

'Gunnery' is the operations of large military guns.

10 Far off, like a dull rumour of some other war.
        What are we doing here?

Emotionally distressing.

The poignant misery of dawn begins to grow ...
We only know war lasts, rain soaks, and clouds sag stormy.

The verb form of the noun 'mass', so meaning 'to gain mass'.

Dawn massing in the east her melancholy army

15 Attacks once more in ranks on shivering ranks of gray,
        But nothing happens.

'dawn' should indicate promise and renewal, but here it just means more misery is to follow. The image is of the German army in its grey uniforms, positioned to the east.

Following one after another.

Sudden successive flights of bullets streak the silence.
Less deadly than the air that shudders black with snow,
With sidelong flowing flakes that flock, pause and renew,

'Nonchalance' means casual or indifferent behavior.

20 We watch them wandering up and down the wind's
        nonchalance,
      But nothing happens.

The snowy weather is more lethal than the bullets.

Key:

Repetition	Half-Rhymes	Alliteration
Assonance	Personification	Imagery

Part II

Pale flakes with lingering stealth come feeling for our faces--

We cringe in holes, back on forgotten dreams, and stare,

25   snow-dazed,

Deep into grassier ditches. So we drowse, sun-dozed,

Littered with blossoms trickling where the blackbird fusses.

         Is it that we are dying?

Slowly our ghosts drag home: glimpsing the sunk fires, glozed

30 With crusted dark-red jewels; crickets jingle there;

For hours the innocent mice rejoice: the house is theirs;

Shutters and doors all closed: on us the doors are closed--

         We turn back to our dying.

Since we believe not otherwise can kind fires burn;

35 Nor ever suns smile true on child, or field, or fruit.

For God's invincible spring our love is made afraid;

Therefore, not loath, we lie out here; therefore were born,

         For love of God seems dying.

To-night, His frost will fasten on this mud and us,

40 Shrivelling many hands and puckering foreheads crisp.

The burying-party, picks and shovels in their shaking grasp,

Pause over half-known faces. All their eyes are ice,

         But nothing happens.

Annotations (left):

'cringe' in this context is a slight movement backwards due to fear.

'fusses' suggests restlessness and concern about something trivial.

'jingle' is a light ringing sound; also a slogan or simple memorable tune. Perhaps an ironic reference to war slogans intended to keep morale high.

Reluctant or unwilling (loath). It is the ultimate structure: the thought that we were born only to die.

Shrivel and pucker both mean wrinkle. Frost, like fire, burns things to a 'crisp'.

Annotations (right):

'stealth' – doing something in a secret, undercover way.

'glozed' is a word that fuses two words - glow and glaze (and possibly gloss).

'our dying' could be read in two ways: a) the men around us who will die; b) our own deaths.

So powerful (invincible) as to never be defeated.

The speaker looks ahead to the night to come and how the intense cold will freeze the dead bodies.

Is he referring to the eyes of the dead, the eyes of the burying party or both?

Poem Overviews 2: Exposure

About the Poem

- The poem was written by Wilfred Owen in 1917.
- The poem is based on his experiences as a soldier on the front line in the trenches in the First World War.
- The winter of 1917 was a particularly bitter, freezing one. The setting is dawn and it is snowing.
- Owen was an officer and his regiment was in a 'salient' (3), one of the most dangerous positions in the front line. He was killed on 4th November 1918, one week before the Armistice was signed.

Ideas, Themes and Issues

- **Waiting and suspense:** The soldiers in the salient suffer because 'nothing happens' (6). In a sense, waiting for conflict is almost worse than the conflict itself. The soldiers have been forced into stasis by the snow and the lack of military conflict. The gunnery far off is like 'some other war' (10), removed from their situation. Owen repeats 'nothing happens' four times in the poem. The repeated line is a continual reminder of the futility of war.
- **Nihilism:** Nihilism is extreme negativity – the belief that nothing in life has any meaning or value. In the sixth verse the poet describes how they dream of home as 'ghosts'. They are back – 'snow-dazed/sun-drowsed' (25, 26). The speaker wonders, 'Is it that we are dying?' (28). It is almost as if death would be a relief from their present torture of waiting. Behind this idea is the fact that we are all, wherever we are, in the process of dying. The verses that follow plunge the speaker into a more nihilistic train of thought. The fires at home, the sun, the harvest, the spring which he once thought 'invincible' (36). Finally, his own faith in God, also seems to be dying. This is what the end of the world feels like – it is his apocalyptic vision.
- **Exposure:** The noun 'exposure' is linked to the **verb** 'to expose' and has several meanings:
  - To leave unprotected (from the weather);
  - To be subjected to danger;
  - to put on display; to suffer and eventually die from the cold. All these meanings can be applied to the poem, which explores mental suffering alongside the physical danger of conflict, death and the loss of belief in any meaning in life.

 Key Point

Nihilism is extreme negativity – the belief that nothing in life has any value or meaning.

Form, Structure and Language

- The poem has a regular rhyme scheme of **half-rhymes** (sometimes called para-rhymes) ABBAC. The rhymes are jarring and not perfect, but rough and uncomfortable, like conditions in the trenches and like the mental strain the men are under.
- The **metre** is jarring and un-pretty. The lines range from 12-14 syllables, which sound slightly jarring. Again, what Owen is describing is a very unnatural situation and the antithesis of being civilised.
- The poem is rich in **alliteration** and **assonance**; colour and sensory imagery assail the reader.
- Repetition is used almost as a refrain: the question that won't go away and can't be answered, and the statement of terrible, world-weary emptiness, 'But nothing happens'.
- First person point of view using 'we' and 'our' gives the reader access to the speaker's thoughts. Owen uses 'we' not 'I', as if he is speaking on behalf of all soldiers.

Quick Test

1. Where is the poem set?
2. What is particularly dangerous about the regiment's position?
3. What do we know about the time of day and the weather conditions?

Key Words

Verb
Half-rhymes
Metre
Alliteration
Assonance

Poem Overviews 2: Storm on the Island

The speaker speaks for a collective, or community, using the pronoun 'we'.

'wizened' means shrivelled with age.

The chorus is a group of singers who sing in unison, commenting on the drama.

Emphasises the emptiness of the place.

'But no' at the start of the line makes it extra emphatic.

'strafes' is to attack with continuous fire or bombs from an aircraft.

We are prepared: we build our houses squat,
Sink walls in rock and roof them with good slate.
This wizened earth has never troubled us
With hay, so, as you see, there are no stacks

5 Or stooks that can be lost. Nor are there trees
Which might prove company when it blows full
Blast: you know what I mean – leaves and branches
Can raise a tragic chorus in a gale
So that you listen to the thing you fear

10 Forgetting that it pummels your house too.
But there are no trees, no natural shelter.
You might think that the sea is company,
Exploding comfortably down on the cliffs
But no: when it begins, the flung spray hits

15 The very windows, spits like a tame cat
Turned savage. We just sit tight while wind dives
And strafes invisibly. Space is a salvo,
We are bombarded with the empty air.
Strange, it is a huge nothing that we fear.

Monosyllables emphasise the solidity of the buildings.

'stacks' and 'stooks' are piles of hay or wheat.

Conversational phrase.

'pummels' is to hit repeatedly, usually with a fist.

An oxymoron – how can something explode 'comfortably'?

'salvo' is a sudden discharge of guns in battle. Here the simile is paradoxical as space (or air) is the weapon: this is emphasised in the final line.

Key:

Simile	Metaphor	Extended Metaphor
Alliteration	Assonance	

About the Poem

- The poem was written by Seamus Heaney and published in 1966.
- An islander describes the community's preparations against the violent storms that ravage the island.
- The island landscape is inhospitable and bleak, and therefore exposed to the storm's intensity and violence, which is powerfully described.
- Stormont in Belfast is the seat of the Northern Ireland Assembly. It has been speculated that the title of the poem is a play on words: (Stormont Ireland) and that the storm's 'huge nothing' (19) is itself a metaphor for Assembly debates.

Ideas, Themes and Issues

- **Nature's power:** The storm gathers pace as the poem develops. At first, it is a tragic chorus and sounds human, though its cry is a lament of sadness. Then, it becomes an explosive force similar to a military attack from enemy aircraft. The poem describes an islander's thoughts on living in an exposed place where they are at the mercy of the storm's natural power. To withstand its onslaught they too must remain strong. The community's power is in the strength of their resistance.
- **Community:** The first six lines describe the community's defence preparations against the fury of the elements. We believe, as the speaker seeks to reassure his listener, that the community will survive and ride out the storm. The speaker's words are calm, reassuring and uplifting, though he twice tells his listener not to expect any 'company' (6, 12) from trees or sea (perhaps a reference to a company of soldiers) to aid them in their resistance. They are alone with their shelters (like air-raid shelters, built low against the landscape). The speaker concentrates on describing how the islanders are capable of withstanding assault. The last line of the poem recalls Franklin D. Roosevelt's rousing words from his inaugural address in 1933 when America was in the Great Depression — the 'only thing we have to fear is fear itself'.

Key Words

Alliteration
Assonance
Extended metaphor
Present tense

Form, Structure and Language

- The poem is written in blank verse with the use of **alliteration** and half-rhyme (squat/slate; cliffs/hits; air/fear) that tie the poem together. The **assonance** intensifies as the storm gathers brutal force.
- The **extended metaphor** is of a military attack. The islanders are a community under attack from the storm, which is likened to a bombing raid.
- The use of **present tense** suggests that the islanders are doing what they have always done and what they always will do – to prepare to weather the storm.

Quick Test

1. Who is speaking in the poem?
2. What events do they describe?
3. Why does the speaker talk of being 'bombarded' with 'empty air' (18)?

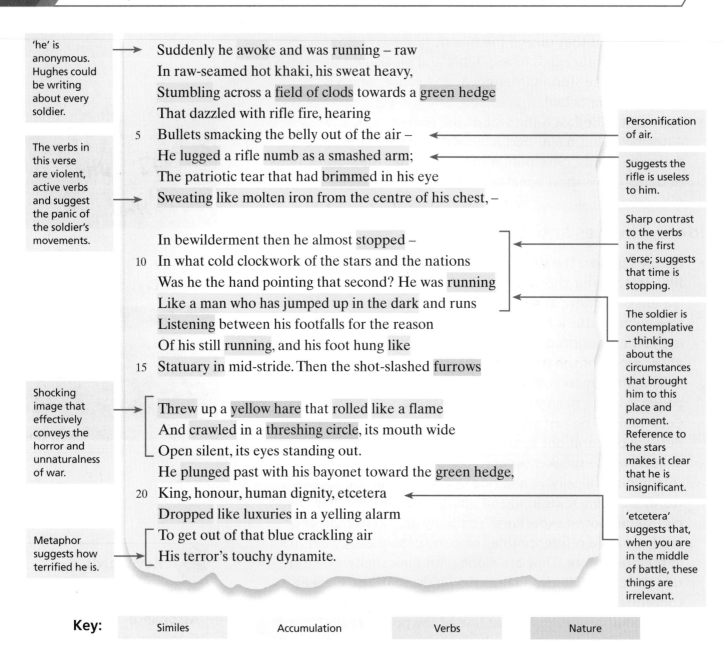

'he' is anonymous. Hughes could be writing about every soldier.

The verbs in this verse are violent, active verbs and suggest the panic of the soldier's movements.

Shocking image that effectively conveys the horror and unnaturalness of war.

Metaphor suggests how terrified he is.

> Suddenly he awoke and was running – raw
> In raw-seamed hot khaki, his sweat heavy,
> Stumbling across a field of clods towards a green hedge
> That dazzled with rifle fire, hearing
> 5   Bullets smacking the belly out of the air –
> He lugged a rifle numb as a smashed arm;
> The patriotic tear that had brimmed in his eye
> Sweating like molten iron from the centre of his chest, –
>
> In bewilderment then he almost stopped –
> 10 In what cold clockwork of the stars and the nations
> Was he the hand pointing that second? He was running
> Like a man who has jumped up in the dark and runs
> Listening between his footfalls for the reason
> Of his still running, and his foot hung like
> 15 Statuary in mid-stride. Then the shot-slashed furrows
>
> Threw up a yellow hare that rolled like a flame
> And crawled in a threshing circle, its mouth wide
> Open silent, its eyes standing out.
> He plunged past with his bayonet toward the green hedge,
> 20 King, honour, human dignity, etcetera
> Dropped like luxuries in a yelling alarm
> To get out of that blue crackling air
> His terror's touchy dynamite.

Personification of air.

Suggests the rifle is useless to him.

Sharp contrast to the verbs in the first verse; suggests that time is stopping.

The soldier is contemplative – thinking about the circumstances that brought him to this place and moment. Reference to the stars makes it clear that he is insignificant.

'etcetera' suggests that, when you are in the middle of battle, these things are irrelevant.

Key: Similes   Accumulation   Verbs   Nature

About the poem

- The poem was written by Ted Hughes and published in his collection *The Hawk in the Rain* in 1957. Hughes's father was a First World War veteran.
- The poem opens with the speaker describing the terrified actions of a soldier as he runs across the battlefield during a bayonet charge.
- For a moment, the soldier thinks about what it is that has brought him to this time and place before a horrifying image of death brings him back to reality.
- The soldier is not named; the poet's choice to make him anonymous makes him represent all soldiers.

Ideas, Themes and Issues

- **Reality of conflict:** The poem vividly describes the events of a conflict, showing their impact on the soldier's mind. The poem is a snapshot of a single moment on battlefield and suggests that the events are so powerful they will never leave the speaker. It is left ambiguous whether it is a real conflict being recalled, if it is a nightmare, or even if it is the waking nightmare of a soldier who has been permanently traumatised by his experience. The soldier feels bewilderment and panic as he charges forward. The poem ends with an image that vividly depicts his terror ('touchy dynamite',(24)).
- **Patriotism:** The poem is more than a description of a single charge in a conflict. It expresses ideas about how conflict affects those involved in it. The poet suggests that the motives that people claim for going to war are luxuries and have no place in the realities of war (20). For this soldier the reasons why he is fighting, and the sense of what he is fighting for, are irrelevant when confronted with the brutal reality of conflict. He started out as a patriotic soldier (7), but that motivation has been transformed into fear ('Sweating like molten iron from the centre of his chest' (8)).

Form, Structure and Language

- **Verbs** give a clear sense of movement in the poem. It feels frantic as the soldier makes his charge. Many of the verbs suggest that he isn't in control of the situation.
- There are only four sentences in the whole poem and **enjambment** adds to the sense of forward movement and his lack of control. There is a lull in the second verse as the soldier stops and contemplates what he is doing there. The phrase 'almost stopped' (9) refers to physically stopping (picked up in line 15 in the simile that likens his foot to a statue) and to mentally stopping to think about the purpose of war.
- The use of **accumulation** in line 20 shows that there are many reasons for going to war, but when a person is in the middle of it they all seem irrelevant.
- The **semantic fields** of war and nature are **juxtaposed**, showing the impact of war on the land.
- **Personification** in line 5 suggests that the air is full of bullets.
- The poem is written in the third person which, along with the fact that the soldier is unnamed, creates a distance between the poet and reader, and the experience described. It also makes the narrative universal, i.e. representative of all soldiers' experiences.

Key Words

Verb
Enjambment
Accumulation
Semantic field
Juxtaposed
Personification

Quick Test

1. How does the poet show that battle is raging around the soldier?
2. What does the poet think about war?
3. How does the poet show his feelings about war?

The Charge of the Light Brigade

1 What is one effect of the relentless rhythm that the poet uses? When does it change during the poem and why?

If feels as if they are truly going to battle. It reflects the uptempo nature approaching, and the confusion upon being fired at.

[3]

2 Why do you think the poet chooses to use the third person for the narrative?

reflects the selfless nature of the army.

[1]

3 In what way does Tennyson use language to suggest the dangers the cavalrymen faced?

repitition. how over the danger is. 'valley of death' N N biblical.

[1]

4 Identify five verbs in the third and fourth verses that suggest the violence of battle.

Charging, plunged, Shattered, Volleys, Stormed.

[5]

5 Pick out one example of strong repetition in each verse. Explain what effect this technique has on the reader.

'half a legue' - rhythmic.

[2]

6 What do you understand by the lines, 'Not though the soldier knew/Someone had blundered' (11-12)?

they didn't know they were going to death.

[1]

7 The word 'wondered' has two meanings. How do you think those meanings fit in the line 'All the world wonder'd' (31 and line 52)?

Wondered why they h.) brave a

[1]

8 Tennyson uses three different images for the valley where the charge took place. Find and list them.

'Valley of death' 'Jaws of death' 'mouth of hell'

[3]

Exposure

1 What do we learn about the condition of the soldiers in the first verse?

They are sat like ducks, cold, but idle. [1]

2 Why does the speaker use the pronouns 'we' and 'our', rather than 'I' and 'my'?

everyone is going through this experience. [1]

3 What is unusual about the speaker's attitude to dawn?

he does not want dawn to come. [1]

4 What does the 'wind's nonchalance' (21) suggest?

the wind has power/ability to be non-chalant. [1]

5 In the fifth verse the speaker suddenly shifts to a different scene. What is the scene and why does he do this?

bringing in bibles adds drama, shows all of life is like this!. [2]

6 What is the speaker's attitude towards home in the sixth verse?

he will only be home dead. [1]

7 What do you understand by the lines, 'For God's invincible spring our love is made afraid;/ Therefore, not loath, we lie out here; therefore were born' (36-37)?

[1]

8 In the final verse, the speaker thinks about the night to come. What does he think will happen?

he will dres, be cold, be killed. [1]

9 The phrase 'but nothing happens' is repeated four times in the poem. What is one effect of this?

anticlimactic, shows futile nature of fighting. [1]

10 Why does the speaker refer only to mice and crickets when he thinks about home?

[1]

Storm on the Island

1 How would you describe the speaker's tone of voice from the words below? Tick the correct answer.

calm ☐ uncertain ☐ cynical ☑ nothing ☐ angry ☐ fearful ☐ **[1]**

2 How have the islanders adapted for the storms?

'root our houses with good slate' **[1]**

3 What is the landscape of the island like?

"bare +barren' earth hasn't troubled us' **[1]**

4 What does the speaker say about trees and the sea?

'Trees and leaves raise a tragic chorus in a gale'
'the sea spits like a tame cat turned savage' **[2]**

5 Pick out three words that are military terms used to describe the storm.

'Pummeled' 'exploding' 'strafes' **[3]**

6 What does the use of the pronouns 'we' and 'our' suggest about the islanders?

they are a community. **[1]**

7 Pick one striking use of alliteration and assonance in the poem. What is the effect?

'tame cat turned savage'
'space is a salvo' **[2]**

8 The storm is an attack of 'nothing' (19). Explain what the poet means by this oxymoron.

they are afraid of empty air. **[1]**

9 Who do you think the speaker is talking to? Why does the speaker try to reassure his listener?

_____

_____

_____

_____

_____ **[2]**

Bayonet Charge

1 The poem starts very abruptly, as if in the middle of a narrative. What is one effect of this?

Feels reflects the narratives actions.

[1]

2 Comment on the poet's use of punctuation in the first verse. How would you describe its effect?

dispunch, recreate runnig over clods.

[2]

3 What do you understand by the simile in line 6?

[1]

4 How does the pace of the poem seem to slow down in the second verse? At what point does the action recommence?

Stopped - physical slop. rhetoric punchuation becomes slow, doubting rather than impulse,

[2]

5 Look again at line 1 and the simile in lines 12–14. What do these lines suggest about the impact this experience had on the soldier?

he begins to question himself

[1]

6 What is your response to the image of the hare in lines 16–18?

[1]

7 'King, honour, human dignity, etcetera' (20). Why are these 'luxuries' to the soldier?

[1]

8 Why do you think has the poet chosen to write this poem using the third person?

[1]

9 How does the poet capture the soldier's sense of rising terror in the final verse?

[1]

10 What happens to the soldier's sense of patriotism in the final lines?

[1]

Ozymandias

1 Four people are mentioned in the poem. Who are they?

Ozymandias, a traveller, the sculptor, subjects. [4]

2 What is one effect of the poet's description of the statue as a collection of parts?

breaks ozymandias's power [1]

3 What do you understand by 'The hand that mock'd them and the heart that fed' (8)?

[2]

4 Which aspects of Ozymandias are described in lines 9–11? How do lines 12-15 present a contrast with this?

[2]

5 What two ideas about power does Shelley encapsulate in the phrase 'colossal wreck' (13)?

[2]

6 The final line contains two examples of alliteration. What is the effect of these?

[1]

7 How does Shelley convey the idea that Ozymandias was a tyrant?

[1]

8 In what ways does a sense of Ozymandias's character remain, even though his statue is a ruin?

[1]

9 'Look on my works, ye Mighty, and despair!' (11) could be read in two contrasting ways. What are they?

[2]

10 What does the description of the once-great ruler suggest about Shelley's view of power and powerful people?

[1]

London

1 The poem contains a list of things the poet hears as he walks through London streets. List four of the things the poet hears.

.... [4]

2 What is the significance of the word 'charter'd' repeated in the first two lines?

.... [1]

3 What impact does the poet's use of the personal voice have on the reader?

.... [1]

4 Is it important that the poet focuses on the plight of young people? Explain.

....

.... [2]

5 One of the poet's intentions was to shock his readers. Pick out one image that fulfils this purpose.

.... [1]

6 Does Blake offer hope, or a sign of future improvement, in the poem?

.... [1]

7 What does the line 'Runs in blood down palace walls.' (12) refer to? Why might it have been relevant to London?

....

.... [2]

8 In the final two lines, references to 'new-born infants' and 'marriage hearse' are juxtaposed. What is Blake suggesting here?

.... [1]

9 The rhyme and rhythm that Blake uses is ballad form, which is a traditional type of song. Why do you think he chooses this form?

....

.... [1]

10 The words 'woe' (4), 'cry' (5) and 'sigh' (11) contain long, heavy vowel sounds. What is one effect of this?

.... [1]

Extract from The Prelude

1 Which poetic form does the poet use?

[1]

2 Which structural devices has the poet used to keep the reader moving on through the recount?

[1]

3 Do you think the boy's reaction to the huge peak was inspired by guilt? Explain.

[2]

4 Pick out a phrase that suggests the boy was determined and purposeful at first.

[1]

5 Which phrase suggests the boy's state of mind on his way back across the lake?

[1]

6 What was the boy's reaction to the huge peak?

[1]

7 What was the boy's reaction to the lake and the moon?

[1]

8 Choose one striking image from the poet's description of the huge peak. Explain its effects.

[2]

9 Describe the effect that his experience had on the boy in the days following.

[1]

10 Why do you think Wordsworth chose to include this particular boyhood memory in 'The Prelude'?

[1]

My Last Duchess

1 The Duke mimics other voices and creates hypothetical situations in his monologue. What does this suggest about his character?

....

.... [2]

2 What reasons does the Duke give for not speaking to the Duchess directly about her behaviour?

....

.... [2]

3 Identify two occasions when the Duke issues commands to the envoy that are disguised as requests.

.... [1]

4 Is it significant that the monologue ends with the word 'me!'? Explain.

.... [1]

5 Comment on the significance of the title. What is suggested by the word 'last'?

.... [2]

6 Why does the Duke keep the painting behind a curtain? Who has access to the painting?

.... [2]

7 In what way does the rhyme scheme conflict with the way that the lines are not end-stopped? What might this suggest about the Duke's character and emotions?

....

.... [2]

8 Why does the Duke's reference to the statue of Neptune suggest about his character? In what ways does Browning draw readers' particular attention to these three lines?

....

....

.... [2]

9 Why do you think the Duke is able to appreciate the picture of the Duchess as he wasn't able to appreciate the reality? What might Browning be suggesting about the relationship between art and reality?

....

....

.... [2]

Starts abruptly, as if in the middle of a conversation or an interview.

'we got sent' tells us the soldier is acting under orders.

'legs it' means to run.

'Well' is a conversational discourse marker.

The soldier makes it clear that the decision was not his alone.

Does he not know the names, or can he not remember? Is he protecting them? Or are the names not important?

A second conversational discourse marker.

Horrifically graphic visual description of the victim's body.

The phrasing and word choice are economical to the point of sounding casual, in stark comparison to the brutality of the shooting.

The soldier is aware it is there, but he manages to ignore it while on patrol.

The image is of the stain left by the victim's blood.

Once he is back home the soldier is haunted by the memory of the incident, that he was able to block out while he was on duty. Alcohol, drugs and even sleep offer no respite from his memories.

'blink' here has two meanings: a) to close and open the eye; b) to waver or have a sudden doubt; here the tone changes.

The soldier feels he cannot remove the stain of blood from his hands. Does he feel personally responsible for what happened?

On another occasion, we got sent out
to tackle looters raiding a bank.
And one of them legs it up the road,
probably armed, possibly not.

5 Well myself and somebody else and somebody else
are all of the same mind,
so all three of us open fire
Three of a kind all letting fly, and I swear

I see every round as it rips through his life –
10 I see broad daylight on the other side.
So we've hit this looter a dozen times
and he's there on the ground, sort of inside out,

pain itself, the image of agony.
One of my mates goes by
15 and tosses his guts back into his body.
Then he's carted off in the back of a lorry.

End of story, except not really.
His blood-shadow stays on the street, and out on patrol
I walk right over it week after week.
20 Then I'm home on leave. But I blink

and he bursts again through the doors of the bank.
Sleep, and he's probably armed, and possibly not.
Dream, and he's torn apart by a dozen rounds.
And the drink and the drugs won't flush him out–

25 he's here in my head when I close my eyes,
dug in behind enemy lines,
not left for dead in some distant, sun-stunned,
        sand-smothered land
or six-feet-under in desert sand,

but near to the knuckle, here and now,
30 his bloody life in my bloody hands.

Key: Half-rhymes   Colloquial language   Verbs of violence   Metaphor

About the Poem

- The poem was written by Simon Armitage and published in the 2008 collection *The Not Dead*.
- The speaker of the poem is a serviceman who has returned from Iraq.
- The man is haunted by his involvement in a shooting which resulted in the violent and bloody death of an Iraqi.
- To write the collection, Armitage interviewed soldiers from the Iraq war, but he uses the first person for this poem, so it is as if the reader is hearing the interview first-hand.

Ideas, Themes and Issues

- **Conflict and its consequences:** The first five verses is a soldier recalling – or reliving – a memory. The poet uses the soldier's speaking voice and **colloquialism** to make the conflict vivid and real for the reader. The phrasing of the sentences mimics real, ordinary and familiar speech ('well' and 'so' are commonly found as connectors in modern everyday conversation). This contrasts starkly with the brutality of the memory recounted. In a sense the poem's structure mimics the soldier's experience, giving us a real understanding of the trauma he is experiencing 'here and now' (29). The poet cleverly uses the experience of reading a poem to re-create the impact of that real-life experience for the speaker. The poem is not about judging the rights or wrongs of conflict. It explores the impact of conflict on one individual.
- **Post-traumatic stress disorder (PTSD):** The poem explores one man's experience of a condition called PTSD. People with this condition continually revisit a traumatic incident in a vivid flashback, but are unable to forget details that they desperately want to forget. The memories they harbour can persist for a lifetime and sufferers of PTSD must learn how to live with them. For this soldier, the act of gunning down the looter, and the sight of his wrecked body, remains vivid in his memory and he re-experiences them continuously – hence the title, 'Remains'.

Form, Structure and Language

- Short clauses, for example in lines 8 and 15, mimic the patterns of natural speech. The language is economical and unadorned, creating the sense that we are reading real-life testimony.
- Verbs of violence, the use of the historic present tense ('legs it', 'are all of the same mind', 'open fire') to narrate past events, together with **colloquial** language and stark imagery, lend immediacy and impact.
- **Half-rhymes** and a bouncy four-beat rhythm create a sense of ironic jollity, disrupted by shorter lines at key moments. The use of **enjambment** creates suspense and also is like real, direct speech.

Key Words

Colloquialism/colloquial
Half-rhymes
Enjambment

Quick Test

1. Who is the speaker in the poem?
2. How does the poet create a sense that this is a real experience?
3. In what way does the victim continue to have 'bloody life' (30)?

Poem Overviews 3: Poppies

Poppies are symbolic of those who have died in war.

Three days before Armistice Sunday
and poppies had already been placed
on individual war graves. Before you left,
I pinned one onto your lapel, crimped petals,
5   spasms of paper red, disrupting a blockade
of yellow bias binding around your blazer.

'crimped' and 'spasms' subvert the poppy's symbolic representation of peace.

Sellotape bandaged around my hand,
I rounded up as many white cat hairs
as I could, smoothed down your shirt's
10 upturned collar, steeled the softening
of my face. I wanted to graze my nose
across the tip of your nose, play at
being Eskimos like we did when
you were little. I resisted the impulse

Alliteration emphasises her attempt to control her feelings.

Metaphor suggests he is too grown up to want signs of affection from his mother.

15 to run my fingers through the gelled
blackthorns of your hair. All my words
flattened, rolled, turned into felt,

Imagery suggests she is turning her emotions inwards.

Dramatic verb 'threw it open' suggests speed and decisiveness.

slowly melting. I was brave, as I walked
with you, to the front door, threw
20 it open, the world overflowing
like a treasure chest. A split second
and you were away, intoxicated.
After you'd gone I went into your bedroom,
released a song bird from its cage.

Suggests his excitement at the experience that awaits him.

25 Later a single dove flew from the pear tree,
and this is where it has led me,
skirting the church yard walls, my stomach busy
making tucks, darts, pleats, hat-less, without
a winter coat or reinforcements of scarf, gloves.

30 On reaching the top of the hill I traced
the inscriptions on the war memorial,
leaned against it like a wishbone.
The dove pulled freely against the sky,
an ornamental stitch. I listened, hoping to hear
35 your playground voice catching on the wind.

'wishbone' may have two meanings: a) she is bent double (bent at the waist in a stooped position) by the force of her loss; b) she wishes he would return safely.

'playground voice' is a reference to his childhood and links to lines 11–14.

Key: Metaphor   Simile   Symbols of remembrance and peace   Maternal love

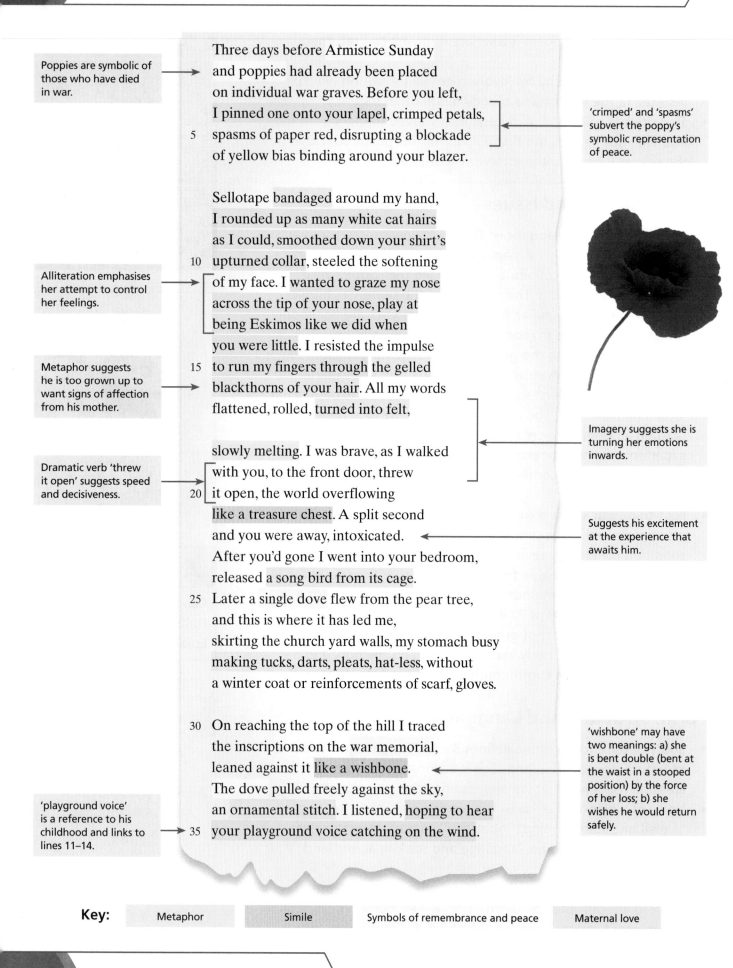

About the Poem

- The poem was written by Jane Weir and commissioned for a collection, *Exit Wounds,* published by *The Guardian* newspaper in 2009.
- A mother describes the day her son left home to join the army. She struggles to contain her emotions as she pins a poppy to his lapel.
- After he has gone she releases her emotions by visiting his bedroom and later visits a local war memorial and remembers him.

Ideas, Themes and Issues

- **Women and conflict:** The poem explores the feelings of a mother left behind and the pain of loss she feels. This reminds the reader that many women have been in her situation in past wars. Mothers, grandmothers, wives and sisters also feel the effects of conflict in their feelings of worry and grief for male relatives, though they may not be directly involved in battle.

- **Ambiguity:** It is not clear whether the speaker's son is one of the dead in the war graves. The opening of the poem could be read as a mother sending her son off to school instead of (as we later realise) to join the army. The mother's pain is made clear (words such as 'bandaged' in line 7; the way she finds it hard to speak in lines 16–18 and the way her stomach is in knots in line 28 show this). We do not know how much time has elapsed between her visit to his bedroom and her walk to the war memorial, or what has happened to him in between. There are several areas of ambiguity in the poem, suggesting that the speaker is struggling to express those same emotions that she fought to conceal from her son. The way that she starts many of her sentences in the middle of the line also indicates how emotional she is feeling.
- Domestic imagery ('sellotape', 'cat hairs') at the start contrasts with the symbols of remembrance and peace ('poppies', 'dove'), and the language of maternal love ('run my fingers through the gelled [hair] (15)) makes it painfully clear how difficult it must be to let a child go to war.

Key Words

Ambiguity
Dramatic monologue
First person
Simile
Metaphor

Form, Structure and Language

- The poem is a **dramatic monologue** written in the **first person**. This form allows the reader into the inner emotions of the speaker. The mother laments the loss of her son. The moment of their parting is a collection of the small things she focused on in order to keep her emotions in check.
- The opening is ominous with the mention of Armistice Sunday and war graves, which are juxtaposed with her son's departure. The use of **similes** (21) and **metaphors** (15) further heighten the emotional response of the reader.

Quick Test

1. How does the poet show feelings about conflict in the poem?
2. What does the poet suggest conflict is like for those left behind?
3. What do the songbird and the dove represent?

Poem Overviews 3: War Photographer

A 'dark room' is where photographers develop their photographs.

Belfast is the capital of Northern Ireland, Beirut is the capital of the Lebanon and Phnom Penh is the capital of Cambodia. All three cities have seen bloody conflict in wars in recent times.

Matter-of-fact statement; as if waking from a reverie, he sets to work.

He is nervous about developing the photographs, though he wasn't when he took them. A reminder that his job is dangerous too and that he needs a steady hand to take good photographs.

Two meanings in the image: a) the subject of the photograph is now dead; b) as the photograph develops it is as a ghostly image before it takes on definition.

We might think he is already in another war zone, but in fact he is talking about England here (he is in a plane above England) 'where ... they do not care'.

A 'spool' is the cylinder on which photographic film is coiled.

Dark rooms are lit by red bulbs during the development process. Duffy likens it to the red tabernacle light found in Catholic churches that symbolises the eternal light of Christ.

A quotation from the Old Testament. It means that human life is not permanent.

Photographs are developed using special fluid 'solutions'. Also, possibly a reference to blood being shed.

His job is important, as it is a record of what is happening; yet he still asks for permission before intruding on grief.

Suggests that the 'reader' is only momentarily moved by the images of war.

'impassively' means without emotion.

In his dark room he is finally alone
with spools of suffering set out in ordered rows.
The only light is red and softly glows,
as though this were a church and he
5   a priest preparing to intone a Mass.
Belfast. Beirut. Phnom Penh. All flesh is grass.

He has a job to do. Solutions slop in trays
beneath his hands, which did not tremble then
though seem to now. Rural England. Home again
10   to ordinary pain which simple weather can dispel,
to fields which don't explode beneath the feet
of running children in a nightmare heat.

Something is happening. A stranger's features
faintly start to twist before his eyes,
15   a half-formed ghost. He remembers the cries
of this man's wife, how he sought approval
without words to do what someone must
and how the blood stained into foreign dust.

A hundred agonies in black-and-white
20   from which his editor will pick out five or six
for Sunday's supplement. The reader's eyeballs prick
with tears between the bath and pre-lunch beers.
From the aeroplane he stares impassively at where
he earns his living and they do not care.

Key: Simile   Metaphor

About the Poem

- The poem was written by Carol Ann Duffy and published in 1985.
- The poem describes the experience of a war photographer, back home in England, developing photographs he has taken and which will be published in a Sunday magazine.
- Written in the third person, Duffy takes us inside the man's thoughts, commenting on the indifference to war by western media and its readers.

Ideas, Themes and Issues

- **Moral dilemmas:** The poem explores the difficulties faced by someone who does a job that records human suffering. The war photographer makes a living out of the horrors he captures on film. It is a dangerous and emotionally draining job. He shuttles between war-torn areas and his home, which is peaceful and removed from that other world, and is struck by the indifference of the former to the latter. He also takes time to adjust to being home; it makes him 'tremble' (8). He knows his job is necessary because people should know about the brutalities of conflict, but at the same time he is aware that his job could be seen as insensitive. He has to intrude on private grief and there is a suggestion that he despises the way he makes money out of suffering. He also dislikes the fact that his photographs are published, not in the main pages, but in the magazine supplement (they are marginalised). In the penultimate lines he stares 'impassively' out of the plane, indicating that he has become emotionally numb. We are left to decide whether the indifference of his English readership is more to blame for this lack of feeling, than the horrors he has seen of war.

- **Western attitudes to foreign conflict:** The poem carries criticisms of western attitudes to foreign conflict and the media's presentation of them. The photographs are not in the main newspaper, but in the 'supplement' (21) that goes with the main newspaper. It can also mean something added or extra. There is an implicit lack of feeling in the way the editor discards a hundred images of death, choosing to publish only five or six.

Form, Structure and Language

- The poem has four verses of six lines each, with a regular rhyme scheme of ABBCDD. Each verse seems like a single frame of a photograph, each revealing a little bit more about the way the photographer feels about his work.
- The poem has religious **imagery** in the use of **simile**. The red developing room light glows as 'though this were a church' (4) (a reference to the red light in Tabernacle Catholic churches) and the photographer is compared to a 'priest preparing to intone a Mass' (5). This underlines that the photographer takes his job seriously, even reverently. The reference to a church ritual is picked up in the final verse with the mention of the Sunday supplement (Sunday being a day when traditionally people go to church).
- The poet uses **metaphor** to give a sense of how the photographs are direct and vivid images of conflict (for example in lines 12 and 19).
- **Contrasts** are made between the war zones where he works ('nightmare heat' (12)) and the peaceful homeland where he develops them ('fields which don't explode beneath the feet' (11)); between the brutality of war ('blood stained into foreign dust' (18)) and the indifference of those who read about it (21–22).
- The second and third, and fifth and sixth lines, are rhyming couplets, suggesting neatness and precision, and recalling the disciplined way the photographer does his job (with his ordered rows of film).

Key Words

Imagery
Simile
Metaphor
Contrasts

Quick Test

1. What is the photographer doing at the start of the poem?
2. Which cities are mentioned as war zones he has travelled to?
3. What does the poet assume about the readers of the Sunday supplement magazine?

About the Poem

- The poem was written by Imtiaz Dharker, who was born in Pakistan and grew up in Glasgow. She is a poet, artist and documentary film-maker. It was first published in Dharker's 2006 collection *The Terrorist at My Table*.
- The poet muses on the fragility of human existence, sparked by an examination of tissue paper.

Ideas, Themes and Issues

- **Fragility and power:** The poet suggests that paper, which is a fragile tissue, has the power to alter and control our existence. Tissue paper alters our view of things; paper is used to record powerful, even sacred, knowledge and information. This leads the poet to speculate that buildings are drawn on thin architect paper; landmarks and geographical features become see-through when they are printed on maps; shop receipts have the power to control our lives, though they are worthless.

- **The power of humans:** The final image is of an architect who creates living human flesh out of all the types of paper that the poem has listed (with their 'shapes that pride can make' (31)) to create something that resembles human skin (another 'tissue'). An architect creates a grand design, using layers of paper from all the different types listed earlier in the poem to create something that is living and human. The poem is saying if we let the light through, like tissue paper does, we see things

The opening image is of tissue paper, thin enough to see through and which alters one's vision of things.

The Koran is the sacred text of Islamic.

'sepia' is a reddish-brown colour, associated with sepia ink, once used for writing.

The verbs 'smoothed' 'stroked' tell us that this paper has the power to evoke love.

Change in mood here. The poet is perhaps thinking of architect's drafts here. Buildings can be marks on paper, too.

Images of someone softly blowing on the paper drafts and scattering them.

This image sees maps as transparent evocations of a concrete world.

Paper that lets the light
shine through, this
is what could alter things.
Paper thinned by age or touching,

5   the kind you find in well-used books,
the back of the Koran, where a hand
has written in the names and histories,
who was born to whom,

the height and weight, who
10 died where and how, on which sepia date,
pages smoothed and stroked and turned
transparent with attention.

If buildings were paper, I might
feel their drift, see how easily
15 they fall away on a sigh, a shift
in the direction of the wind.

Maps too. The sun shines through
their borderlines, the marks
that rivers make, roads,
20 railtracks, mountainfolds,

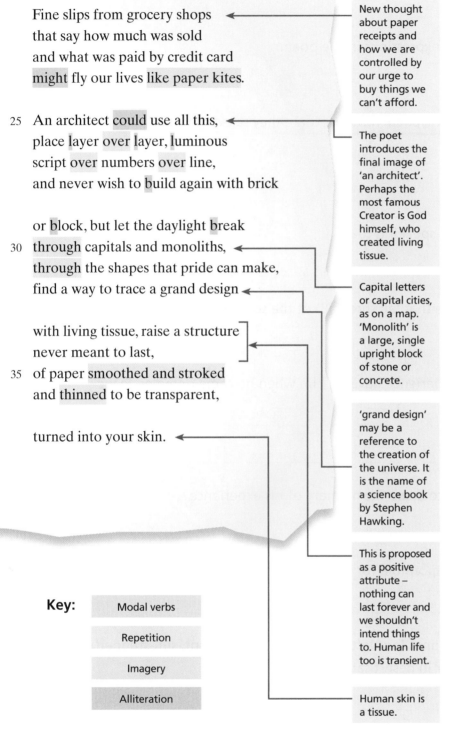

Fine slips from grocery shops
that say how much was sold
and what was paid by credit card
might fly our lives like paper kites.

25 An architect could use all this,
place layer over layer, luminous
script over numbers over line,
and never wish to build again with brick

or block, but let the daylight break
30 through capitals and monoliths,
through the shapes that pride can make,
find a way to trace a grand design

with living tissue, raise a structure
never meant to last,
35 of paper smoothed and stroked
and thinned to be transparent,

turned into your skin.

New thought about paper receipts and how we are controlled by our urge to buy things we can't afford.

The poet introduces the final image of 'an architect'. Perhaps the most famous Creator is God himself, who created living tissue.

Capital letters or capital cities, as on a map. 'Monolith' is a large, single upright block of stone or concrete.

'grand design' may be a reference to the creation of the universe. It is the name of a science book by Stephen Hawking.

This is proposed as a positive attribute – nothing can last forever and we shouldn't intend things to. Human life too is transient.

Human skin is a tissue.

Key:
- Modal verbs
- Repetition
- Imagery
- Alliteration

differently and we can be free to live. The poet urges us to consider that things we might think of as permanent are not, in fact, any more permanent than our own human tissue.

Form, Structure and Language

- The poet uses **imagery** of types of paper to explore ideas about solidity and certainty, change and impermanence in buildings, countries, borders and the landscape.
- The poet uses 'everyday', almost mundane language, heavily patterned with **assonance** and **alliteration**, to convey complex, intricate ideas and images.
- **Modal verbs** express uncertainty and possibility. Like tissue paper, the poem's language mystifies. Though the words themselves are easy to understand, their subtle meanings are elusive.
- **Repetition** is used to imitate the layering of paper referred to in the seventh verse. Though the lines in each verse have a regular line length, and appear as blocks on the page, the poet subverts this regularity and creates a layering effect by use of **enjambment** across lines and even verse boundaries (8–9, 28–29, 32–33).

Key Words

Imagery
Assonance
Alliteration
Modal verbs
Repetition
Enjambment

Quick Test

1. Which object begins the poet's train of thought?
2. What types of paper are mentioned?
3. What type of 'tissue' is specifically mentioned?

Remains

1 Find two examples of colloquial expressions in the poem.

.... [2]

2 What meanings does the title of the poem have? How are they relevant to the content of the poem?

....

.... [2]

3 Who do you think is the speaker speaking to?

.... [1]

4 Why does the speaker emphasise that 'all three' (8) of the soldiers shot at the looter?

.... [1]

5 What effects did the soldier's experience have on him when he came home on leave?

....

....

.... [2]

6 What does the soldier do to try to erase the memory of his experience?

.... [1]

7 Do you think the soldier feels responsible for what happened? Explain.

.... [1]

8 What do you understand by the lines 'not left for dead in some distant, sun-stunned, sand-smothered land/or six-feet-under in desert sand' (27–28)?

.... [1]

9 In what ways does the structure of the poem mirror the soldier's experience?

....

.... [1]

10 How does the poet's use of tenses contribute to the impact of the soldier's recount?

.... [1]

Poppies

1 Who is the speaking voice in the poem?

[1]

2 Why might the mention of poppies at the start of the poem be significant?

[1]

3 Why does the mother resist the impulse to rub noses with her son? What does this suggest about her emotions?

[2]

4 Is the mother hopeful that her son will return? Explain.

[2]

5 What do you understand by the simile 'like a wishbone' (32)?

[1]

6 How does the speaker remind us of the child that her son once was?

[2]

7 How does the speaker describe her words to her son? What does this suggest about her recollection of the moment?

[2]

8 Find two places in the poem where the speaker experiences emotional conflict.

[2]

9 The poem contrasts domestic words and phrases related to details of war. Find two domestic details in the poem.

[2]

War Photographer

1 What is the image in the first verse? What does it suggest about the photographer's attitude towards his job?

.... [1]

2 How does his homeland offer him a contrast to the places he goes for his job?

.... [1]

3 How does the photograph in the third verse affect him?

.... [1]

4 In what way does the speaker suggest the impersonality of the magazine editor?

.... [1]

5 What is suggested by the phrase 'between the bath and pre-lunch beers' (22)?

.... [1]

6 Where is the photographer in the final two lines?

.... [1]

7 What is your interpretation of the final two lines?

....

.... [1]

8 Do you think the newspaper editor also has a difficult job? Explain your answer.

....

.... [2]

9 What do the lines: '... his hands which did not tremble then/though seem to now' (8–9) suggest?

.... [1]

10 What is the effect of the image 'a half-formed ghost' (15)? Give two possible meanings.

.... [2]

Tissue

1. What is the mood of the poem? Tick the correct answer.

   joyful ☐   sad ☐   reflective ☐   angry ☐   [1]

2. Paper is used as a metaphor in the poem. What things might it represent?

   .... [1]

3. The poet contrasts images of strength and permanence against the fragility of paper. Name two such images from the poem and the concepts they represent.

   .... [2]

4. Why does the poet think an architect might prefer paper over 'brick or / block'?

   .... [1]

5. What do you think the poet means by the phrase 'a grand design' (32)?

   .... [1]

6. Why is it important that what an architect makes is 'a structure/never meant to last' (33–34)?

   .... [1]

7. What materials does the 'architect' in the seventh, eighth and ninth verses work with?

   ....

   .... [1]

8. In what ways are the poet's mentions of light and air important in the poem?

   .... [1]

9. The poet places the last line on its own, set apart from the verses. Why do you think she does this?

   .... [1]

10. Several of the types of paper in the poem are used to record measurements. What does this suggest to you?

    ....

    .... [1]

The Charge of the Light Brigade

1 Why do you think Tennyson chose to portray this military disaster as a triumph of bravery?

_____ [1]

2 Find a line in the poem that implies a criticism of the decision to charge.

_____ [1]

3 Tennyson uses rhetorical questions twice in the poem. Find them and analyse their effects.

_____

_____

_____

_____ [3]

4 Why do you think the speaker chooses to focus on the charge rather than the context surrounding it?

_____

_____ [1]

5 Comment on the poet's use of rhythm.

_____

_____

_____

_____ [3]

6 Identify three examples of the poet's use of emotive language.

_____

_____

_____ [3]

7 What is one effect of the highly formal, rhythmical and repetitive structure of the poem?

_____ [1]

8 Which aspects of conflict do you think are emphasised by the poet?

_____ [1]

9 Find a quotation that indicates this charge was instantly famous.

_____ [1]

Exposure

1 In what ways do the meanings of the word in the title of the poem reflect its content?

_____

_____ [1]

2 How does the poet use alliteration and assonance to achieve effects in the first verse? What effects are created?

_____

_____ [2]

3 What does the poet's use of metre suggest about the emotional state of the soldiers?

_____ [1]

4 What are two effects of the repeated line 'but nothing happens'?

_____ [1]

5 Part I of the poem sets the scene of the soldiers' situation. What ideas does Part II develop?

_____

_____ [1]

6 Why does the speaker think that the snow is more deadly than enemy fire?

_____ [1]

7 What image does the speaker use to describe the fires of home? Why is it significant?

_____

_____ [2]

8 'knife us' (1) and 'nervous' (4) are half rhymes. Why does the poet use half-rhymes in this poem?

_____ [1]

9 What is powerful about the phrase, 'All their eyes are ice.' (39)?

_____

_____ [1]

Storm on the Island

1 How is nature presented in the poem? .... [1]

2 Select two phrases where the speaker directly addresses the listener/reader. What is the effect of this conversational tone?

....

.... [2]

3 Why does the poet choose the phrase 'exploding comfortably' (13) to describe the sea's actions?

....

.... [1]

4 What is the extended metaphor that the poet uses in the second half of the poem?

.... [1]

5 How are the islanders presented in the poem? Choose from the following:

passive   **sad**   **brave**   **resigned**   **terrified**   **resilient**   **angry** [1]

6 What does the speaker realise in the final line?

.... [1]

7 In what ways does the speaker create a sense of the islander community in the poem?

.... [1]

8 What do the references to 'company' (6, 12) mean in the poem?

....

.... [2]

9 The speaker lists several negatives in the poem and is aware of the dangers presented by the storm. Is his overall tone positive or negative?

.... [1]

10 Are wider political connotations suggested by the situation in the poem? What other kinds of 'empty air' (18) are we 'bombarded' (18) with?

....

.... [1]

Bayonet Charge

1 What is the effect of starting the poem with the word, 'Suddenly'?

.... [1]

2 How does the language in the third line seem to enact a sense of 'stumbling' through mud?

.... [1]

3 What is unusual about the image used to describe the hedge?

.... [1]

4 Which two emotions are juxtaposed at the end of the first verse?

.... [2]

5 Find an example of enjambment that conveys the sense that the soldier 'almost stopped' (9) going forwards.

.... [1]

6 What is one effect of juxtaposing nature with the violence of battle?

.... [1]

7 The soldier asks himself a question in the second verse. In your own words, explain what he is questioning at that moment.

.... [1]

8 What prompts the soldier to continue his forward charge in the third verse?

.... [1]

9 Comment on the poet's use of imagery in the final two lines.

....

.... [1]

10 In what ways could the poem be described as a description of a recurring traumatic memory?

....

.... [1]

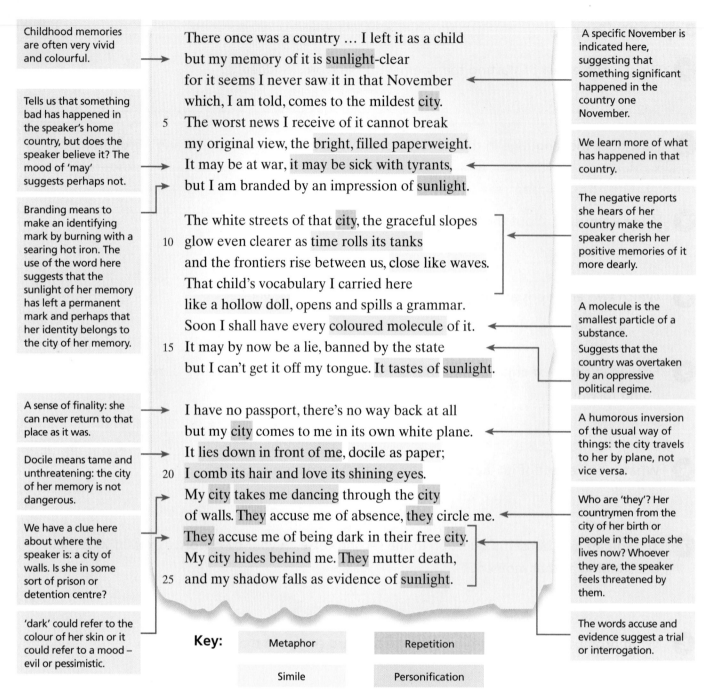

Childhood memories are often very vivid and colourful.

A specific November is indicated here, suggesting that something significant happened in the country one November.

Tells us that something bad has happened in the speaker's home country, but does the speaker believe it? The mood of 'may' suggests perhaps not.

We learn more of what has happened in that country.

Branding means to make an identifying mark by burning with a searing hot iron. The use of the word here suggests that the sunlight of her memory has left a permanent mark and perhaps that her identity belongs to the city of her memory.

The negative reports she hears of her country make the speaker cherish her positive memories of it more dearly.

A molecule is the smallest particle of a substance.
Suggests that the country was overtaken by an oppressive political regime.

A sense of finality: she can never return to that place as it was.

A humorous inversion of the usual way of things: the city travels to her by plane, not vice versa.

Docile means tame and unthreatening: the city of her memory is not dangerous.

We have a clue here about where the speaker is: a city of walls. Is she in some sort of prison or detention centre?

Who are 'they'? Her countrymen from the city of her birth or people in the place she lives now? Whoever they are, the speaker feels threatened by them.

'dark' could refer to the colour of her skin or it could refer to a mood – evil or pessimistic.

The words accuse and evidence suggest a trial or interrogation.

There once was a country ... I left it as a child
but my memory of it is sunlight-clear
for it seems I never saw it in that November
which, I am told, comes to the mildest city.
5   The worst news I receive of it cannot break
my original view, the bright, filled paperweight.
It may be at war, it may be sick with tyrants,
but I am branded by an impression of sunlight.

The white streets of that city, the graceful slopes
10   glow even clearer as time rolls its tanks
and the frontiers rise between us, close like waves.
That child's vocabulary I carried here
like a hollow doll, opens and spills a grammar.
Soon I shall have every coloured molecule of it.
15   It may by now be a lie, banned by the state
but I can't get it off my tongue. It tastes of sunlight.

I have no passport, there's no way back at all
but my city comes to me in its own white plane.
It lies down in front of me, docile as paper;
20   I comb its hair and love its shining eyes.
My city takes me dancing through the city
of walls. They accuse me of absence, they circle me.
They accuse me of being dark in their free city.
My city hides behind me. They mutter death,
25   and my shadow falls as evidence of sunlight.

Key: Metaphor   Repetition

Simile   Personification

About the Poem

- The poem was written by Carol Rumens and published in 1993.
- The poem explores the effects on a person when their own language and identity are threatened. The speaker remembers the city of her childhood that she left behind and explores how her memory of it continues to sustain her.
- No specific country or city is mentioned and these may be a **metaphor** for a forbidden relationship or lover. The idea of being an emigrée could be an extended metaphor to explore different kinds of emigration and the concept of having to leave a place or person.

Ideas, Themes and Issues

- **The power of memory:** The speaker left the country of her birth when she was a child, but her memories of the city are so vivid that they are conveyed in the present tense using images of light, white and brightness. In her memory, the city continues exactly as it was, even though her country of birth is no longer as she remembers. The power of memory to become a source of strength is further explored in the third verse, when the speaker's imagination brings the city to her and gives it life. The pronoun changes from 'that city' to 'my city', showing her personal feeling about it. It provides her with a defence of 'sunlight' against the serious threats of death that she faces from 'them'.

- **Displacement and loss:** The speaker has been displaced from her own country and her reaction to this is to cling harder to the things she has lost – her language and identity. The phrase 'I have no passport' (17) suggests she has not adopted a new nationality and 'there's no way back at all' (17) tells us of the forced absence from the country of her birth. We start to wonder if this poem is multi-layered: could the city be a metaphor for faith, religion or even for love? The speaker talks of her city as coming to help her in a white plane. We think of the kind of paper plane children make and then imagine it as something with hair and eyes (like a pet, or a doll), as a child would. Her city-memory, though fragile and docile, represents personal freedom ('takes me dancing through the 'city of walls' (21–22)) and personal power.

- **Opposition:** The poem is full of contrasts and oppositions: the 'then' of the speaker's memory versus the 'now'; full of threats and violence; the speaker versus 'them'; and the city of her birth versus the one she lives in now, which is a 'city of walls' (21–22).

Form, Structure and Language

- The poem is written as a **soliloquy**, with the speaker musing as if to herself.
- Longer, more lyrical sentences describe the memory-city and contrast with shorter, terser sentences used for the speaker's current, difficult situation.
- Metaphors are used to explore different types of isolation or absence linked to the meaning of the title, *Emigrée* ('frontiers rise between us' (11); 'time rolls its tanks' (10), 'My city takes me dancing through the city/of walls' (21–22)).
- **Repetition** of 'sunlight' shows that light is very important to the narrator as a symbol of freedom. The repetition of 'city' shows the importance of her childhood memory to her now, as an emigrée.
- Repetition of the unnamed 'they' (22–24) suggests menace and oppression.

Quick Test

1. When did the speaker leave the country of her birth?
2. What has happened to the city since she left it?
3. Why does the speaker need to draw on her memories now?

About the Poem

- The poem was written by Beatrice Garland and published in 2007.
- In the poem, a Japanese mother explains to her children what happened to their grandfather in the war. The poem gives a female perspective on conflict.
- The mother imagines what her father may have thought about his previous life as a fisherman while he is in the air.
- She describes the consequences of her father's decision to abort the mission.
- Kamikaze were suicide attacks made by Japanese airmen during the latter stages of the Second World War. Aircraft were deliberately crashed into military targets, usually warships.

Ideas, Themes and Issues

- **Choice and decisions during conflict:** The speaker's father had a choice to live or die and in undertaking the kamikaze mission during the conflict of the Second World War he had chosen death. On the way, he changes his mind. However, the poem makes it clear that the choice he made was between honourable suicide and living with dishonour, ostracised by society.
- **Four generations:** There is both a **parallel** and a **contrast** between the pilot's mission with its elaborate ceremonial preparations and his father's

Samurai were highly skilled feudal Japanese warriors, who followed the code of Bushido: loyalty and honour until death. Some kamikaze pilots took their sword with them, if they were officers.

A magic spell or chant. Ceremonies, such as sharing a cup of sake or water, were carried out before kamikaze pilots departed on their final mission.

Marks a change in his feelings.

The speaker doesn't know for sure what went through her father's mind.

A stack of stones piled together.

Waves that break into foam.

A parallel is drawn here with the pilot's father, who also went on dangerous fishing missions at sea and returned safe.

Her father embarked at sunrise
with a flask of water, a samurai sword
in the cockpit, a shaven head
full of powerful incantations
5 and enough fuel for a one-way
journey into history

but half way there, she thought,
recounting it later to her children,
he must have looked far down
10 at the little fishing boats
strung out like bunting
on a green-blue translucent sea

and beneath them, arcing in swathes
like a huge flag waved first one way
15 then the other in a figure of eight,
the dark shoals of fishes
flashing silver as their bellies
swivelled towards the sun

and remembered how he
20 and his brothers waiting on the shore
built cairns of pearl-grey pebbles
to see whose withstood longest
the turbulent inrush of breakers
bringing their father's boat safe

Key:

| Colour imagery |
| Similes |
| Metaphor |
| Alliteration |
| Repetition |

25 *- yes, grandfather's boat – safe*
to the shore, salt-sodden, awash
with cloud-marked mackerel,
black crabs, feathery prawns,
the loose silver of whitebait and once
30 *a tuna, the dark prince, muscular, dangerous.*

And though he came back
my mother never spoke again
in his presence, nor did she meet his eyes
and the neighbours too, they treated him
35 *as though he no longer existed,*
only we children still chattered and laughed

Expresses the social disgrace of not carrying out his mission.

till gradually we too learned
to be silent, to live as though
he had never returned, that this
40 *was no longer the father we loved.*
And sometimes, she said, he must have wondered
which had been the better way to die.

Her father's experience changed his and their lives forever.

We never hear what the speaker's father thought, as he never spoke of it.

life as a fisherman who was also engaged in dangerous missions, but returned safe. The pilot recalls how his brothers would wait for his father's return, which may have prompted him to think about his own children, who would not have seen him return from this mission if he had not turned back.

- **Judgement:** The speaker is careful not to judge her father's decision, nor to offer excuses. She speculates about his reasons for turning back and not completing his mission, but leaves the readers to make up their own mind about his decision. Only in the final two lines does the speaker suggest that her father might not have been happy with the choice he made.

Poem Overviews 4: Kamikaze

Form, Structure and Language

- The poet uses italics to show the mother's **direct speech**, which adds impact to her words.
- The colour imagery used to describe nature suggests the vibrancy of the life that the pilot did not want to lose.
- **Metaphors** and **similes** create vivid images of the beauty, energy and freedom of ocean life.
- The poet uses repetition (three of the verses start with the word 'and') and **connectives** to show the narrative flow of the speaker. The concluding sentence also starts with 'and', connecting the speaker's train of thought to what has gone before.
- **Alliteration** of *f, sh* and *s* sounds in the third verse creates a sense of energy and freedom in the natural world, but there are also echoes of the military mission at hand. The mention of the sun perhaps suggests the Japanese flag.

Quick Test

1. What was the 'one-way journey into history' (5)?
2. Why was the speaker's father treated as though he no longer existed?
3. Why aren't we told what the speaker's father actually thought about his experience?

Key Words

Parallel
Contrast
Direct speech
Metaphor
Simile
Connectives
Alliteration

Poem Overviews 4: Checking Out Me History

About the Poem

- The poem was written by John Agard, a black British poet who was originally from Guyana in the Caribbean, and was published in 2007.
- In the poem, Agard reclaims black identity and makes the reader aware that history is only a point of view.
- Agard reminds us that control of the past means control of the present.
- The poem explores the relationship between culture, power and identity, and the importance of finding our own history and voice.

Ideas, Themes and Issues

- **Power:** Agard powerfully reclaims his own history in this poem, taking back the power that he feels was taken from him by the history he was taught. The figures from white history are carefully chosen to mix real people (Napoleon) and storybook figures (man in the moon) and those who are the stuff of myth (Robin Hood). This both undermines the power, and questions the authenticity, of the conventional history he has been taught. By the end of the poem, Agard has wrestled power back, doing the 'checking' (52) of his own history and actively 'carving' (53) his own identity.
- **Contrast and metaphor:** In the longer italicised verses, the speaker's tone changes to respect and reverence as he talks about the famous black figures from history. His language here also contrasts with that used in the quatrains. The longer verses are rich in positive imagery and metaphor, 'See-far woman' (27) and he uses unusual and original language. The image 'yellow sunrise' (48), suggests that Seacole was full of hope and promise for the soldiers she helped. The structure of the poem cleverly contrasts official and non-official history, represented by the two types of verse.

- **Anger:** The speaker is clearly angered by being prevented from knowing about his own history, shown in the **quatrains** that start 'Dem'. 'Dem' is repeated 17 times in the poem and like the beat of a drum, or a hammer, this repetition powerfully expresses the speaker's outrage.

Form, Structure and Language

- **Non-standard English** is used to show the speaker's own culture and difference from the history of white culture taught to him.
- A repeated quatrain is used, each with a different famous white figure from history, followed by a contrasting black figure from history, who has been ignored.
- Rhyme in the repeated quatrains gives a slightly mocking tone; **free verse** in the longer verses, which detail black historical figure, are more serious in tone.
- Agard purposefully uses **enjambment** (as in 'A slave/ With vision/Lick back/Napoleon' (11–14)) to replicate the rhythm of natural speech and his lack of **punctuation** represents his rejection of white history.
- Repetition is a prominent feature of the poem, used as an expression of the speaker's powerful emotions.

Them (perhaps the speaker's history teachers). →

Dem tell me
Dem tell me
What dem want to tell me

'Bandage' and 'Blind' suggest his view (of history and himself) has been distorted. →

Bandage up me eye with me own history
5 Blind me to me own identity

An important date in English history which marked the takeover of English society by the French. →

Dem tell me bout 1066 and all dat
dem tell me bout Dick Whittington and he cat

A poor boy with nothing but a cat who became Lord Mayor of London – mentioned in fairy tales and pantomimes. →

But Toussaint L'Overture
no dem never tell me bout dat

10 *Toussaint*
a slave
with vision
lick back
Napoleon

The black leader of the Haitian revolution against French Colonial rule. He abolished slavery in Haiti and turned it into an independent republic. →

15 *battalion*
and first black
Republic born
Toussaint de thorn

Italics make these verses stand out visually, setting them apart from the 'white history'. →

to de French
20 *Toussaint de beacon*
of de Haitian Revolution

Dem tell me bout de man who discover de balloon
and de cow who jump over de moon

Reference to a nursery rhyme 'Hey Diddle Diddle'. →

Dem tell me bout de dish ran away with de spoon

Maroon is a term used for black slaves who escaped from slavery and lived in remote places within the areas they had been transported to. Nanny de Maroon led the maroons to victory in Jamaica against British rule. →

25 but dem never tell me bout Nanny de maroon

Key:

Repetition	Non-standard English
Rhyme	Metaphor

Nanny
see-far woman
of mountain dream
fire-woman struggle
30 *hopeful stream*
to freedom river

Dem tell me bout Lord Nelson and
Waterloo

but dem never tell me bout Shaka de
great Zulu
Dem tell me bout Columbus and 1492
35 but what happen to de Caribs and de
Arawaks too

Dem tell me bout Florence Nightingale
and she lamp
and how Robin Hood used to camp
Dem tell me bout ole King Cole was a
merry ole soul
but dem never tell me bout Mary Seacole

40 *From Jamaica*
she travel far
to the Crimean War
she volunteer to go
and even when de British said no
45 *she still brave the Russian snow*
a healing star
among the wounded
a yellow sunrise
to the dying

50 Dem tell me
Dem tell me wha dem want to tell me
But now I checking out me own history
I carving out me identity

A famous English admiral who inflicted a massive defeat on the French which meant the British Empire was strengthened.

The Battle of Waterloo was the final defeat of the French army led by Napoleon which made Britain a super power.

An influential leader who united the Zulu people against other tribes and the increasing presence of the white settlers.

The original inhabitants of the West Indies at the time when Columbus 'discovered' the islands. They were killed off by the war or through exposure to European diseases.

A British nurse famous for working in the Crimean War and her innovations in hygiene.

Legendary English hero who supposedly stole from the rich to give to the poor.

A Jamaican-born woman who provided food and lodging for convalescent officers in the Crimean War.

A conflict between the Russian Empire and Britain, France and the Ottoman Empire (1854-56). It resulted in many deaths from fighting and disease.

Quick Test

1. Who does the speaker refer to as 'dem'?
2. What is the effect of the repetition in the opening verse?
3. What do 'bandage' (4) and 'blind' (5) suggest about the speaker's view of the history he has been taught?

Key Words

Quatrain
Non-standard English
Free verse
Enjambment
Punctuation

The Emigrée

1 What *might* the city symbolise? Explain your answer.

.... [1]

2 Pick out two images that suggest war or conflict in the second verse.

....

.... [2]

3 The speaker's memory of the city is described using various images. Choose two and explain their effects.

....

.... [2]

4 What does the line, 'I have no passport ...' (17) suggest about the speaker's situation?

.... [1]

5 Why do you think she can never return to the city of her childhood?

.... [1]

6 How does the speaker keep the memory of her city alive?

.... [1]

7 How would you describe the speaker's attitude and tone?

.... [1]

8 Who do you think 'they' are (third verse)?

.... [1]

9 In the third verse, the speaker talks of the city hiding behind her. In what way might she protect the city?

....

.... [1]

10 Does the speaker believe the bad reports she hears of her home-city? Choose a quotation to support your answer.

.... [1]

Kamikaze

1 What preparations did the pilot make before he started his mission?

_____ [1]

2 What, according to the speaker, did he see first from the plane?

_____ [1]

3 The phrase 'must have' is repeated. What does it tell us about the poem?

_____ [1]

4 In the third verse, what does the poet's use of imagery and colour suggest?

_____ [1]

5 Why do you think it is significant that the pilot might have thought about his brothers?

_____ [1]

6 What did the pilot's father do for a living?

_____ [1]

7 Why, in the speaker's view, did her father decide to abort his mission? Give two reasons.

_____

_____ [2]

8 In what ways did the speaker's family live as though 'he had never returned' (39)?

_____ [1]

9 Do you think the speaker thinks her father was right to abort the mission?

_____

_____ [1]

10 Explain what you understand by the final two lines.

_____

_____ [2]

Checking Out Me History

1 What is the speaker's tone in the quatrains of the poem?

.... [1]

2 How does the speaker's tone change in the longer, italicised verses?

.... [2]

3 What do the references to fairy stories add to the speaker's view of the history he has been taught?

.... [1]

4 In what ways does non-standard English create a distinctive voice in the poem?

.... [1]

5 What are the effects of the use of repetition in the poem?

....

.... [1]

6 How does the tone of the poem change at the end? Which words indicate this?

....

.... [2]

7 'Dem' is repeated 17 times in the poem. What is the effect of this repetition?

....

.... [1]

8 Name two figures from black history that the speaker admires. How does he express his admiration?

....

.... [3]

9 What did those figures from history possess that the speaker, by the end of the poem, is claiming for himself?

.... [1]

Remains

1 What were the soldiers told to do at the start?

.... [1]

2 What do you think the soldier found most shocking about the shooting?

.... [1]

3 What do you think the soldier means when he says, ' Well myself and somebody else and somebody else/are all of the same mind' (5)?

....

....

.... [1]

4 Comment on the choice of the word 'Remains' as a title for the poem.

....

....

.... [1]

5 How does the poet use verbs to convey the violence of the shooting?

....

.... [2]

6 The shooting is described twice in the poem. Explain why you think the poet does this.

.... [1]

7 Does the speaker's sense of responsibility for the killing shift as the poem develops? Identify quotations to support your view.

....

....

.... [1]

8 What condition is the soldier suffering from? .... [1]

9 Does the poem invite readers to feel sympathy for the speaker or for his victim? Explain.

....

....

.... [2]

Poppies

1. The speaker says that she 'was brave' (18). What was she brave about?

    _____ [1]

2. Identify one detail that shows the speaker hid her real emotions from her son.

    _____ [1]

3. Which image suggests a sudden release of the speaker's emotions?

    _____ [1]

4. Find two specific words, or phrases, that express the speaker's emotional pain as if it were physical.

    _____ [2]

5. Explain what you understand by the simile, 'the world overflowing/like a treasure chest' (20–21).

    _____ [1]

6. The mother takes care that her son doesn't feel like the child that for her, he always will be. Why do you think she does this?

    _____ [1]

7. Why do you think the poet chooses to include references to needlework and embroidery in the poem?

    _____ [1]

8. How does the speaker seek solace from the pain of separation?

    _____ [1]

9. Do you think the speaker remains in control of her emotions once her son has gone? Explain.

    _____

    _____ [1]

10. Describe the emotions that you think the speaker is feeling as she leans against the war memorial.

    _____ [1]

War Photographer

1 What is the effect of comparing the photographer to a priest in the first verse?

.... [1]

2 How does the structure of the poem contribute to what the poet wants to say about the war photographer?

....

....

....

.... [4]

3 How does the war photographer feel about his job?

....

.... [2]

4 What does the line, 'how he sought approval/without words to do what someone must' (16–17) suggest?

.... [1]

5 What possible two meanings are there in the sentence 'Something is happening.' (13)?

....

.... [2]

6 In the third verse, the images 'twist' before the photographer's eyes. What two possible meanings are suggested?

.... [1]

7 Why do you think the photographer is 'impassive' in the final two lines?

.... [1]

8 Do you find the final two lines of the poem positive and uplifting or negative and downbeat? Explain your answer.

....

.... [2]

Tissue

1 What sort of paper begins the speaker's train of thought? _____ [1]

2 What two qualities does she say the paper has?

_____ [2]

3 Though paper is thin and fragile, how does the poet argue that it has power?

_____ [1]

4 In what three ways does the poet suggest the insubstantiality of paper?

_____ [3]

5 In the eighth verse, what does the phrase 'break through' and the repeated word 'through' suggest about the architect's task?

_____

_____ [2]

6 Why do you think the poet chooses to refer to 'an architect' as the creator in the seventh, eighth and ninth verses?

_____ [1]

7 Which type of paper contains a criticism of commercialism in society? Find the phrase that suggests this.

_____

_____ [2]

8 How would you describe the attitude of the speaker? Is she optimistic, pessimistic, idealistic, fanciful or something else? Explain your answer.

_____

_____ [2]

9 In what ways are the structure and language of the poem reflective of and indicative of its content?

_____

_____ [1]

The Emigrée

1 How does the speaker retain contact with her home city?

.... [1]

2 Why is it important to her that she keeps the memory of her city alive?

.... [1]

3 Why do you think the speaker's language is 'banned by the state' (15)?

.... [1]

4 Why does the city of her memory have to travel to the speaker?

.... [1]

5 How does the poet use repetition to suggest the speaker's feelings about her past?

....

.... [1]

6 What is the effect of the repeated word 'sunlight' at the end of each verse?

.... [1]

7 Explain what you understand by the use of the word 'dark' in the third verse.

.... [1]

8 Why does the speaker imagine her city as a doll in the third verse?

....

.... [2]

9 What do you understand about the speaker's situation in the third verse?

....

.... [2]

10 What do you think the city might symbolise?

.... [1]

Kamikaze

1 What do we learn about the pilot in the first verse?

.... [1]

2 Do you think it is possible the pilot saw fishes from his plane? Why does the speaker imagine he did?

....

.... [2]

3 What memory does the speaker think the shoals of fish sparked in her grandfather?

.... [1]

4 What is the significance of the speaker's grandfather's job?

.... [1]

5 In what ways did the speaker's father's decision change his and the family's life forever?

.... [1]

6 In what way are the fishing boats that the speaker's father might have seen connected to his decision to abort his mission?

....

.... [1]

7 Why do you think the poet chooses imagery of colour and light to describe the fish in the sea and the catch in the third and fifth verses?

.... [1]

8 What is the effect of the final, short sentence?

....

.... [1]

9 The poet uses italics to show the speaker's direct speech. What is the effect of this in the poem?

.... [1]

Checking Out Me History

1 What are the main themes of the poem?

[3]

2 How does the poet show the contrast between 'white' and 'black' history through the structure of the poem?

[1]

3 Why does the poet choose to use no punctuation in the poem?

[1]

4 What is one effect of the repetition of 'dem tell me'?

[1]

5 Which nursery rhyme is referred to in the poem? [1]

6 Who was Mary Seacole? Which famous person does the poet compare her to?

[2]

7 Where in the poem does the poet choose to use metaphor? Why does he make this choice?

[2]

8 Name two folk history figures that the speaker references. What are the effects of including these figures?

[2]

9 Why does the poet choose to use non-standard English in the poem?

[1]

10 Which lines tell you that the speaker is taking back power over his own identity?

[1]

Poem Analysis: Comparing Poems

Understanding the Question

- In the exam you will be asked to compare two poems from the anthology cluster you have studied.
- The questions will ask you to compare a named poem, which will be printed on the paper, with another poem of your choice from the same cluster.
- A list of all the poems in the cluster will be printed for you as a reminder, but you will not have them in front of you.
- Therefore you must make sure you know all the poems really well so you can write a good comparison answer.
- Your answer needs to include:
  - Your understanding and response to the poems.
  - Textual references (quotations or paraphrases) that support your interpretations.
  - Appropriate use of subject terminology.
  - Analysis of the language, form and structure of the poems.
  - Analysis of the poets' methods and how they use them to achieve meanings and effects.
  - Reference to and understanding of the relationships between poems and the contexts in which they were written.

Key Point

Remember that you will only have the named poem in front of you, so make sure you know all the poems really well.

- The question will usually start with the word 'Compare' and a poem from the cluster will be specified in the question. You must use this as one of the poems in your comparison. You then need to choose another poem from the cluster for your comparison.
- Make sure the poem you choose for your second comparison poem is suitable for the exam question.
- You will have no choice of question, so it is important that the poem you choose has enough similarity and difference with the given poem, and is appropriate for the question.
- For example, look at this question:

> Compare the ways that poets present ideas about the effects of conflict in 'Exposure' and one other poem from 'Power and Conflict'.

- You will always be asked to write about the poets' **presentation**. This means you must show understanding of the poets' methods: structure, language, choice, techniques, tone and attitude.
- Be aware that you can talk about the ideas and theme of the poem, but this alone will not earn you high marks.
- Examiners are looking for a sophisticated **analysis** and exploration of poetic methods and their effects.
- Students who offer different personal interpretations will also be highly rewarded.

Using Comparison Grids

- When you are revising, it may be helpful to use comparison grids to organise your ideas about the poems side by side. Decide on two poems to compare at a time.
- The grid might look like this:

Question focus: e.g. Poets' presentation of the the effect of conflict.		
	Poem 1	**Poem 2**
Themes and ideas – similar – different		
Structure and form – similar – different		
Language features (imagery, repetition, interesting word choices, alliteration, etc.) – similar – different		
Tone and attitude – similar – different		
Personal response and interpretations		

Key Point

Using a comparison grid can be especially helpful when revising the poems.

- Remember that your comparison must consider similarities and differences. You do not need to talk about both of these equally, but you must always be comparing both poems.
- A very common error is to talk about one poem in detail and make only passing mention of the other poem. Plan well so you have plenty to write about on both poems.
- In the exam you may prefer to **annotate** (underline or highlight) parts of the poem given in the question with comparison points for your second poem written alongside.
- This can work well, but it's a good idea to use a comparison grid similar to the one above when you are practising comparison answers, to help you get into the habit of organising your ideas under the right headings.

Key Point

It is vital that you remember to write about both poems, comparing their similarities and differences in relation to the question. Do not talk about just one poem in detail.

Key Words

Presentation
Analysis
Annotate

Poem Analysis: Comparing Poems

Planning

- In the exam you should allow yourself 10–12 minutes for gathering your ideas and writing your plan.
- Here is a template of a plan:

Introduction	Always state the titles (and poets) that you intend to compare. Always mention the focus of the question to show you are aware of it. Show that you are setting out to write in a comparative way.
Main paragraphs	Include at least three paragraphs. You should have at least three clear points of comparison to make – more, if you have more ideas! Make sure you include references to the poems and use quotations to support your points.
Conclusion	Sum up your main ideas, mention the question focus again and give a personal response to the poems.

- Your plan might look like this:

Introduction	Presentation of results of conflict: *Exposure* and *Remains*. Similarities – both show how conflict affects a psychological change on soldiers. The use of first person (singular in *Remains*; plural in Exposure) gives a sense of immediacy and impact to both poems.
Main paragraphs	1. Similar themes/attitudes: the speakers have both been directly involved in conflict. The setting of *Exposure* is at the Front in the First World War during a lull in the conflict and describes the tension of waiting for action; in *Remains* the action of killing is recalled by the speaker as an on-going memory using the historic present tense and he goes on to describe the after-effects of that one incident. 2. Language and effects: Owen's poem is formally divided into two parts – part I describes their situation at the Front, where 'nothing happens'; part II muses on the future they do not have and the lack of belief they have in anything as a result of their war experiences ('for love of God seems dying').

 Key Point

Planning your answer is important, so spend the recommended time creating your plan.

	Remains also can be split into two parts at the words 'But I blink', the second part replaying the first. Use of imagery to show results of conflict: Armitage's 'blood-shadow' is a startling reminder of the violence and its longstanding effect on the soldier; Owen has 'crusted dark-red jewels' a reminder of bloody violence in the fires at home: the result of conflict is to blight everything they once believed in. Note the use of half-rhyme in both poems ('silent/salient'; 'week/blink') to create sense of jarring and unnatural situations that are the results of conflict. Relevance of the titles in each poem, both refer to the results of conflict on the speaker. 3. Context: Armitage's speaker – the use of modern colloquial language ('letting fly', 'carted off') juxtaposed with vivid images ('sun-stunned') are punchy and powerful talking about a recent but distant war; Owen's despair and suffering on behalf of all soldiers describes a world-view shared by a generation after 1918.
Conclusion	Both show the results of conflict as wholly negative: destroys belief in *Exposure* and causes lasting psychological damage in *Remains*.

Writing Your Answer

- When it comes to writing a good comparative essay, there are certain words and phrases that will come in very useful as a way of organising your thoughts to make it easy for the examiner to follow your writing.
- Some useful phrases are:
  - 'Both ... however' – this helps you to mention points of similarity followed by difference. For example,

  > *'Both poets use rhyme to suggest control, however in Browning's poem it is disturbed by the use of enjambment, suggesting the Duke's struggle to control his emotions ...'.*

  - 'Although ... in ...' – you could use this when you want to highlight a difference or contrast between two poems. For example,

  > *'Although Tennyson uses rhythm to evoke the sound of the galloping horses' hooves, in Agard's poem the driving rhythm emphasises the strength of the speaker's emotions.'*

- Other useful phrases to make links are shown alongside:

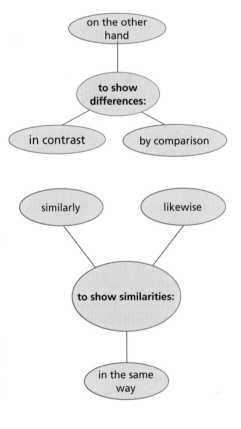

Poem Analysis:
Sample Worked Question and Answer

Compare how the results of conflict are presented in 'Exposure' and in **one** other poem from 'Power and Conflict'. *(30 marks)*

In 'Exposure' and 'Remains', we see the results of conflict on the mental state of individuals. 'Exposure' shows how conflict affects a soldier in the trenches and how it makes him contemplate life and question the reason for existence. 'Remains' shows how the results of conflict can live with a soldier after he has returned home. Both poems reveal how experience of conflict leaves permanent psychological scars.

> Clear introduction which tells the examiner which poems you will cover and refers closely to the question.

Armitage uses the structure of 'Remains' to convey how the memory of a single event replays in one man's mind. The first five verses recount the shooting, described in horrific detail: 'every round as it rips through his life', 'pain itself, the image of agony'. The poem turns in the final line of the fifth verse, 'But I blink' and then repeats the events of the shooting, this time as a waking nightmare sequence with words and phrases repeated from the first part, 'rounds', 'probably armed, possibly not'. The soldier cannot get rid of the memory; drugs and drink won't help him forget. Owen divides 'Exposure' into two parts and the second builds on the situation described in the first. In the first part, the soldiers are waiting for the enemy to attack as dawn approaches. The suspense builds, 'But nothing happens'. Owen repeats this line to convey the tension of waiting, but also war's futility. In the second part he develops this theme, allowing his thoughts to develop and consider the earlier question, 'What are we doing here?'

> Refers to poet's methods with textual references.

> Good focus on structure as a point of comparison.

> Good use of embedded quotation.

> Thoughtful, referenced response showing insightful understanding of the text.

Owen's use of half-rhyme creates a disjointed and unnatural feeling in the reader, conveying the unease and discomfort of the waiting soldiers. His world view is entirely nihilistic at this point. He no longer believes in anything and the images he uses of home, 'sunk fires glozed', 'crusted dark-red jewels' and closed doors are beautifully evocative, but negative. The embers of the fire remind us of wounds. In that cold winter night he can no longer believe that the spring will come. His words speak of his utter despair, 'For love of God seems dying'. His experience of conflict has destroyed his belief in existence. This time when he repeats the line 'But nothing happens' we know he is referring to life itself, as well as the futility of the war and the soldiers awaiting attack.

> Analysis of writer's methods with subject terminology used judiciously; exploration of writer's methods on reader.

> Shows excellent understanding of the poem.

> Develops and explores the effects of writer's methods on reader. Offers an interpretation using a critical style.

Armitage also uses half-rhyme (agony/by/body/lorry) to create a sense of disharmony and unease in the reader. He uses line length to disrupt the rhythm and draw the reader's attention to crucial turning points, 'End of story, except not really' marks where the soldier starts to live with the after effects of the shooting. Again, 'not left for dead in some distant, sun-stunned, sand-smothered land' emphasises that the victim is still alive in the soldier's head.

> Critical, exploratory comparison; judicious use of precise references to support interpretation.

> Judicious use of precise references to support and illustrate interpretation.

Both poets use first-person speakers (one singular; one plural). Armitage's use of the soldier's own words with the casual colloquialisms and idioms ('tosses his guts', 'carted off') are very typical of how an ordinary young man might speak. This makes us empathise with the man's suffering in the final three verses. Here the speaker's language changes into a more poetic and artful mode 'sun-stunned, sand-smothered'. There is a reference to Lady Macbeth's blood-stained hands that won't wash clean, that suggests the soldier's feeling of guilt that won't go away. Owen's use of the first person is very powerful and speaks collectively on behalf of all soldiers. Conflict has left Owen with a profound sense of pointlessness. It has left Armitage's soldier with an unending feeling of guilt, shame and horror.

> Point, evidence, analysis.

> Comparative exploration of ideas/perspectives/contextual factors shown by specific, detailed links between context/text/task.

Both poems have titles chosen because they have multiple relevant meanings. 'Remains' refers to all that is left of the shooter's corpse (his remains) and the way that he remains indelibly in the soldier's mind. 'Exposure' refers to the soldiers at the front who are exposed to death whether from the icy cold or enemy fire, and Owen himself exposes his own despair and mental suffering alongside the physical danger and discomfort.

> Critical, exploratory comparison of the two poems.

Overall, I think both poets show results of war as wholly negative, but in different ways. War is shown to destroy belief in 'Exposure'. It is shown to cause psychological damage in 'Remains'.

> Clear personal response is given.

> Clear conclusion that refers back to the question.

Mark Scheme

- For the anthology question, your answer will be assessed on three Assessment Objectives (AOs). AO1 and AO2 are worth up to 12 marks and AO3 is worth up to 6 marks.
    - AO1 - critical, exploratory comparison; judicious use of precise references to support interpretations
    - AO2 - Analysis of writers' methods with subject terminology used judiciously; exploration of effects of writer's methods on reader
    - AO3 - Exploration of ideas/perspectives/contextual factors shown by specific, detailed links between context/text/task

Key Point

The mark scheme descriptors are available on the AQA website and are worth reading to give you an idea of what the examiner will be looking for.

Examiner's comment

- This is a Level 6 response. It is a strong answer which focuses on detail in an analytical manner and makes very clear links between the poems.
- The answer is focused and demonstrates clear understanding of both poems. The comparison clearly explains the effect of writers' methods in language, structure and form.
- This student uses subject terminology appropriately to support the points made. Different interpretations are explored, including reference to contextual aspects, clearly related to the texts and the specific question asked.

Mixed Exam-Style Questions

Note: In the exam, each question will only refer to one poem. You will need to choose the other poem. For the purposes of these exam-style questions, a poem has been suggested, (in brackets) to make the comparison with. This is shown in the answer section at the back of the book.

1 *Ozymandias* (Compare with *My Last Duchess*)
Compare how poets present ideas about the effects of power in 'Ozymandias' and in **one** other poem from 'Power and Conflict'.
Continue your answer on a separate piece of paper.

30 marks

2 *London* (Compare with *The Emigrée*)
Compare the ways poets present feelings about power in 'London' and **one** other poem from 'Power and Conflict'.
Continue your answer on a separate piece of paper.

30 marks

3 *The Emigrée* (Compare with *Poppies*)
Compare the ways poets present feelings about separation because of conflict in 'The Emigrée' and **one** other poem from 'Power and Conflict'.
Continue your answer on a separate piece of paper.

30 marks

4 *Tissue* (Compare with *Checking Out Me History*)
Compare how poets present attitudes to personal power and identity in 'Tissue' and in **one** other poem from 'Power and conflict'.
Continue your answer on a separate piece of paper.

30 marks

5 *Bayonet Charge* (Compare with *Exposure*)

Compare how poets present attitudes to warfare in 'Bayonet Charge' and in **one** other poem from 'Power and Conflict'.
Continue your answer on a separate piece of paper.

30 marks

6 *The Charge of the Light Brigade* (Compare with *Bayonet Charge*)
Compare the methods poets use to explore ideas about patriotism in 'The Charge of the Light Brigade' and in **one** other poem from 'Power and Conflict'.
Continue your answer on a separate piece of paper.

30 marks

7 ***Exposure*** (Compare with ***War Photographer***)

Compare the methods poets use to explore ideas about the futility of war in 'Exposure' and in **one** other poem from 'Power and conflict'.

Continue your answer on a separate piece of paper.

30 marks

8 ***War Photographer*** (Compare with ***Remains***)

Compare how poets explore ideas about the effects of war in 'War Photographer' and in **one** other poem from 'Power and conflict'.

Continue your answer on a separate piece of paper.

30 marks

9 ***Kamikaze*** (Compare with ***Poppies***)

Compare the ways poets present the impact of conflict on families in 'Kamikaze' and in **one** other poem from 'Power and Conflict'.

Continue your answer on a separate piece of paper.

30 marks

10 *Poppies* (Compare with *War Photographer*)

Compare the methods poets use to explore feelings about war in 'Poppies' and in **one** other poem from 'Power and Conflict'.

Continue your answer on a separate piece of paper.

30 marks

11 *Remains* (Compare with *Bayonet Charge*)

Compare how poets explore ideas about the effects of violent conflict or individuals in 'Remains' and in **one** other poem from 'Power and Conflict'.

Continue your answer on a separate piece of paper.

30 marks

12 *Storm on the Island* (Compare with *Ozymandias*)

Compare the methods poets use to explore ideas about power or powerlessness in 'Storm on the Island' and in **one** other poem from 'Power and Conflict'.

Continue your answer on a separate piece of paper.

30 marks

13 *My Last Duchess* (Compare with *Checking Out Me History*)

Compare the methods poets use to present attitudes towards power in 'My Last Duchess' and in **one** other poem from 'Power and Conflict'.

Continue your answer on a separate piece of paper.

30 marks

14 *Checking Out Me History* (Compare with *Extract from The Prelude*)

Compare the methods poets use to present attitudes towards a sense of giving personal identity in 'Checking Out Me History' and **one** other poem from 'Power and Conflict'.

Continue your answer on a separate piece of paper.

30 marks

15 *Extract from The Prelude* (Compare with *Storm on the Island*)

Compare how poets show the effects of nature's power in 'Extract from The Prelude' and in **one** other poem from 'Power and Conflict'.

Continue your answer on a separate piece of paper.

30 marks

Answers

Page 9 Quick Test Answers

1. The poem is set in an unspecified 'antique land' (1) to make it clear that Ozymandias's achievements belong in the past. The place where the statue is found has no name because nothing of the great city remains. We learn about the statue third hand, creating distance between us and Ozymandias.
2. That it is fleeting. Nothing of Ozymandias's empire remains, but a half-buried statue.
3. Through details of his expression: 'frown' (4) and 'sneer' (5); the contempt with which the artist view him and the arrogance of his self-given title, 'king of kings' (10).

Page 11 Quick Test Answers

1. That the people have no power (everything is 'charter'd' (7)) and they do not seem to be able to change their thinking so that they can make a change.
2. Everything in the poem is negative. Words and phrases are repeated which suggests everything is bleak. The metre of the poem reinforces this idea.
3. To make people think about the society they lived in and realise they could change it.

Page 13 Quick Test Answers

1. At the cove of a lake.
2. It is evening.
3. He immediately took it for a trip on the lake.

Page 15 Quick Test Answers

1. The poem flows from one line to the next, moving on as the Duke is moving on with his next marriage.
2. Hearing the Duchess's fate from the Duke's mouth emphasises his coldness and lack of remorse. It is shocking to us that he sees no shame in telling the servant. He assumes his behaviour will be condoned, but we are horrified by it.
3. To emphasise the fact that the murder of his wife does not affect his conscience. She is now totally under his control because she is another object to boast about, just like the bronze Neptune.

Ozymandias

1. The narrator, who we assume is the poet, is speaking. (He recounts the travellers tale.)
2. Sonnet.
3. 'wrinkled lip and sneer of cold command' (5) suggests he was a heartless ruler.
4. 'Ruler of nothing'; ironic because that is

what he has become now (though it is not what he was).
5. He admires the sculptor more; for the way the sculptor's art has captured its subject's character so perfectly.
6. That the traveller has returned from an ancient civilisation.
7. Ironic because Ozymandias's 'works' (11) referred to in his inscription must have once existed, in what is now a vast area of desert.
8. The poet, the traveller, Ozymandias; They give us three perspectives and create a sense of distance between the reader and what happened to Ozymandias's kingdom. They give a mythical quality to the narrative.
9. Through the ruins of his statue and the sculptor's artistry (and, of course, through the artistry of Shelley's sonnet).
10. It could refer to other powerful rulers: those who lived at the same time as Ozymandias and those to come afterwards.

London

1. A negative and depressing picture of the unremitting misery of its inhabitants.
2. He is depressed by what he sees, (but also angry).
3. It emphasises that he sees nothing, but misery everywhere he looks. There is no hope anywhere.
4. The rhymes carry the reader along; OR the repetitive rhythm reflects the monotonous lives that Londoners lead.
5. Mention of chimney sweeps, the church, soldiers and the palace. Every single strata of society is mentioned.
6. It is literally blackened by the factory smoke of industrialisation; a second meaning is that is morally corrupt because it does nothing to help the people.
7. Example: 'hapless soldier' (11) is interesting because it has two meanings: 1) the soldier is doomed to die; 2) the soldier has no agency to change his state: he acts only for the state. Blake is criticising war here.
8. Not directly, but there is implicit criticism of both the Church and State.
9. To elicit the reader's sympathy and to increase the impact of his shocking imagery.
10. The peoples' own minds are oppressed (in 'manacles' (8)). They are not free and their imaginations are suppressed.

Extract from The Prelude

1. He says he was led there by nature/ 'her' (7).
2. 'It was an act of stealth' (5). Later references are: 'stole my way' and 'covert' (30, 31).
3. Yes, because we are not sure if the fact that his mind was 'troubled' (6)

makes him more susceptible to the impact of the huge peak. It is as though the peak chastises him for his misdemeanour. (Also, he rows excitedly partly because he knows he is doing something wrong.)
4. He is exhilarated by the adventure and enjoys the beauty of the moon and lake.
5. He fixes his eyes on the horizon.
6. It is suggested by 'Went heaving through the water like a swan'(20).
7. 'When, from behind that craggy steep …' (21).
8. As the boat nears the peak it seems to get bigger, due to the change in the boy's perspective. In the boy's imagination, it appears to grow and then move as if it were alive.
9. He was subdued and chastened.
10. The boy's view of nature has changed as a result of his experience; Instead of seeing nature as pleasant and restful, he now sees it as a powerful and awe-inspiring, even dangerous, force.

My Last Duchess

1. A servant of the Count; He has come to negotiate the marriage of the Duke to the Count's daughter.
2. Not the first/Are you (12-3)
3. That the Duchess was kind and considerate to everyone and was positive about everything in her life.
4. (a) '(since none puts by/The curtain I have drawn for you, but I)' (9–10); (b) 'as if she ranked/My gift of a nine-hundred-years-old name/With anybody's gift.' (32–34).
5. It highlights the cold, emotionless and sinister way he used his power to have the Duchess Killed.
6. His self-interruptions ('how shall I say?' (22)); and use of elision (''twas all one!' (25); 'E'en then' (41)). They emphasise that this is an informal conversation (though with a diplomatic backdrop) and give the impression that the Duke is at ease and off-guard (though of course he isn't really – he is giving an elaborate performance).
7. She liked everything – she wasn't the discriminating snob that he wanted her to be. (He implies that she was unfaithful though this may well be a figment of his jealous imagination.)
8. It suggests, ironically, that he is still obsessed by her, even after her death, (but that obsession has transferred to the painting as an artefact).
9. 'Looking as if she were alive' (2); 'and there she stands (4).
10. Dramatic monologue: 1) the speaker is clearly distinct from the poet; 2) an audience is suggested, but never appears in the poem; 3) the revelation of the Duke's character is the poem's primary aim.

Pages 21 Quick Test Answers

1. He writes a poem about them, which suggests he feels their achievement should be celebrated. Words like noble and hero are clearly positive. He reflects on their sense of patriotism, which was considered very important at the time.
2. Onomatopoeia and rhythm suggest the sounds of battle, which seems terrifying. 'Cannon' repeated shows they were surrounded. The valley is personified and also called 'Hell'. 'Not the six hundred' (38) shows that they gave their lives for their country.
3. It suggests that the soldiers' lives were cut short, but it also sounds like an epitaph for them – they will be remembered. It uses positive language ('honour', 'glory') to make it clear that he thinks they should be remembered for ever as heroes.

Page 25 Quick Test Answers

1. At the front line in the trenches during World War I.
2. They are on the front line and surrounded on three sides by the enemy.
3. It is some time before dawn and it is bitterly cold and snowing.

Page 27 Quick Test Answers

1. An islander.
2. The preparations the islanders make to protect themselves against storms.
3. He describes the storm as an attack by enemy aircraft. Instead of bombs, the storm attacks with wind.

Page 29 Quick Test Answers

1. The poem opens in the middle of the battle. The verbs used show that he is moving fast to avoid being killed. The description of the way he 'lugged' (6) his gun shows how he is struggling to get away from danger. The shocking image of the hare shows that nothing is safe.
2. He thinks it is terrifying and that people involved in war simply do what they are told without thinking. If they stop and think about it, they are terrified too.
3. Though images of fear and death. He thinks that people have reasons for going to war, but these are not convincing on the battlefield, when nothing matters but survival.

The Charge of the Light Brigade

1. It mimics the sound of horses' hooves in the charge; it perhaps suggests the sound of a beating heart and the unquestioning and courageous way the cavalry obeyed the order to charge; The rhythm changes in important lines in the poem such as 'All the world wonder'd' (31) and 'Rode the six hundred' (17) (echoed in 'Noble six hundred' (55)).
2. The third person makes the poem sound official, not personal – almost like a newspaper report. It also gives the sense that the cavalrymen were not individuals. The six hundred acted as one.
3. He uses repetition: 'Cannon' in the third and fifth verses; 'Flash'd' in the fourth verse to indicate the strength of the enemy's position, surrounding the cavalry on three sides and armed with guns. Tennyson also refers to the valley of Death' in the first and second verse.
4. Five from: volley'd, thunder'd, plunged, sabring, reeled, storm'd, shatter'd, sunder'd.
5. Repetition of 'Half' (first verse), 'Theirs' (second verse), 'Cannon' (third verse), 'Flash'd' (fourth verse), 'Cannon' (fifth verse), 'Honour' (sixth verse). It gives a sense of the relentless forward motion of the charge and the repetitive sound of the horses' hooves in the charge. Cannon is used twice as a repeated word to show how the cavalrymen went into the valley, turned, and came out again, enduring the cannon-fire twice.
6. The line tells us that Tennyson was aware the order to charge was the result of a misunderstanding by the commanders. (The soldiers knew it would be a suicide mission, but they carried out the order anyway.)
7. It can mean 'to admire or marvel' and it can mean 'to have doubt'. It is possible that Tennyson meant both meanings to be at play: 'the world' (31) probably did admire the cavalrymen's bravery, but it might also have questioned the soundness of the strategy to charge.
8. Valley of Death; jaws of Death; mouth of Hell.

Exposure

1. They are exhausted, but forcing themselves to stay awake (there is no fighting, but they are waiting for it to start).
2. He is speaking collectively for the regiment.
3. Dawn usually signals the promise of a new day and a new start, but for the speaker it only signals more 'melancholy' (13) – sadness – to come.
4. That the wind is indifferent and uncaring of the soldiers' suffering.
5. A rural scene in springtime; It is part-dream and part-hallucination to suggest the confusion in the soldiers' minds: they don't really know where they are any more and desperately try to imagine themselves in a beautiful, restful place.
6. That home as they knew it is closed to them and empty. They will not go there except as ghosts.
7. One interpretation is the spring that we thought will always come, we now fear, will never come. Not even God's power can make it come. There is nothing left to live for and because everything we love has died, we are not unwilling to die here.
8. Exactly the same will happen tomorrow night as happened tonight: (dead bodies will freeze and will be collected for burial).
9. The phrase emphasises that nothing ever

changes: the soldiers are waiting a German attack, but it doesn't come. (It also underlines the futility of the war in that it has achieved nothing and solved nothing.)
10. To emphasise the emptiness of the house – no family is there to welcome the soldiers' home. (It further underlines the soldier's lack of belief in the war because they do not think they are fighting for anything, and have nothing to return home to.)

Storm on the Island

1. calm
2. They have built the houses low to the ground, with deep foundations and sturdy slate roofs.
3. It is bleak and inhospitable. It has no fertile earth. Crops and trees do not grow there.
4. Neither offers any aid or support; there are no trees on the island to offer resistance to the storm and the sea turns wild against the islanders during a storm.
5. 'strafes' (17), 'salvo' (17) and 'bombarded' (18).
6. That they are a community, made close-knit by their shared resistance against the storms.
7. Assonance of 'i' and alliteration of 's' and 't' in lines 14–16 mimics the sound of the storm, and the savage, spitting cat that the wind has been likened to.
8. He means that the storm consists of nothing more than wind, which is empty air.
9. He may be speaking to someone new to the island, as he explains how the islanders have adapted to living with storms. It could be that the person is nervous of the storms, or he may just be curious. The speaker does not underestimate the power of the storms, but he sounds almost proud of how the islanders withstand them.

Bayonet Charge

1. It makes it very personal and engages the reader immediately.
2. He uses commas and dashes to indicate shorter and longer pauses; These give a sense of the breathless, panicked urgency that the soldier feels.
3. It suggests that the rifle is useless to him.
4. The staccato nature of the punctuation disappears with a long question followed by a long sentence and a simile, slowing down the pace as the soldier suddenly thinks about and questions what he is doing as he almost stops. After he sees the hare the action restarts with 'He plunged' in the third verse.
5. Line 1 suggests this is might be a recurring nightmare ('he awoke') and this is echoed in the simile 'Like a man who has jumped up in the dark ...' (12)
6. It is a horrifying image of the violence of war and its impact on nature and innocent animals. (It prompts us to make a comparison between the hare and the soldier.)
7. They are all irrelevant to the soldier at this point.

8. Perhaps because it describes something that was not a personal experience for him – it is imagined. Another idea is that the soldier himself remembers the experience in the third person in order to distance himself from it.
9. By the end all he can see is 'blue crackling air' (22) and all he is aware of is his own terror.
10. It ceases to exist, in the face of his own terror.

Pages 34-37 Review Questions

Ozymandias

1. The poet/speaker; the traveller; Ozymandias; the sculptor.
2. The statue is a ruin – only pieces of it remain and nothing is left of the kingdom that Ozymandias once ruled. (It is a metaphor for the transience of power and influence.)
3. It is a tricky line to make sense of because of the syntax Shelley uses. 'The hand' may be that of the sculptor and 'mock'd' may mean copied , so it could mean that the sculptor traced the features of Ozymandias well; 'mock'd' has a further meaning – scorned –perhaps suggesting that the sculptor managed to convey his own scornful attitude towards his subject in his work. Or the line may refer to Ozymandias's passions (6) and the fact that, though he was scornful of his subjects, he also provided for them.
4. His arrogance (indicated by his boastful and challenging inscription) in lines 9–11; His power was temporary (indicated by the vast sandy desert surrounding the ruined statue) in the final lines.
5. 'colossal' suggests huge power and influence; 'wreck' shows that the power and influence has decayed, lying in ruins.
6. It highlights the vast barrenness and emptiness of the desert and suggests the futility of human power and influence compared to nature's enduring power.
7. By his sneering expression of superiority and 'cold command' (5) which suggests heartlessness.
8. Through the sculptor's ability to capture his character in Ozymandias' facial expressions.
9. It could be seen as a challenge and a boast. It could also be viewed (ironically) as a warning of how power and influence are merely temporary.
10. That power does not last as nature does; that powerful people are egotists; more concerned with their own image and fame than in doing good for others.

London

1. Four from: cries of men; fearful cries of infants; cries of chimney-sweeps; sighs of soldiers; curses of young harlots.
2. The people themselves own and control nothing – not the streets or buildings and not even the river itself.
3. It makes it more immediate and real. We feel we are present in London with the poet as he walks around, hearing and

observing everything. (We are more inclined to believe him.)
4. Yes, because the plight of young people makes it more shocking for the reader; and implies that this situation of misery and oppression will continue to the next generation.
5. Example: 'youthful harlot's curse' (14) is shocking because it suggests that young women, or even girls, are working as prostitutes, and they use profane language to wish ill on everyone.
6. Absolutely not.
7. The violent executions of the French monarchy during the French Revolution; It sparked unrest in London and there was a genuine fear that the same thing might happen in England.
8. That future generations will be corrupted by what is happening now.
9. Traditional ballads are narrative poems and are sentimental or romantic in tone, often telling stories about folk heroes. Here Blake uses the form as a kind of protest song about a city he cares for.
10. The sounds mimic the sounds of misery that Blake hears as he goes through the streets.

Extract from The Prelude

1. Blank verse: unrhymed iambic pentameter.
2. Enjambment; conjunctions.
3. It was at least partly inspired by a guilty conscience; also the fact that he was alone and because it was getting dark: these things together caused his imagination to play tricks on him.
4. One from: 'like one who rows/Proud of his skill'(11–12); 'lustily/I dipped my oars into the silent lake' (17–18); 'Went heaving through the water like a swan' (20).
5. The phrase 'with trembling oars' (29) (suggests he feels upset and frightened).
6. At first he panics and rows faster, but that makes it worse as the peak seems to grow. Then, he decides to turn back, but his arms are trembling with fright.
7. He enjoys the beauty of the lake, the moon and the way the light plays on the water.
8. Two striking images are: 'with voluntary power instinct/ Upreared its head'; (23–24) 'with purpose of its own/And measured motion like a living thing/ Strode after me' (27–29). The narrator, who we assume is the poet, is speaking. He recounts the traveller's tale. Both images personify the huge peak, endowing it with the ability to move in a menacing way.
9. He feels depressed by the understanding of nature that the experience gave to him. As a result of the incident he is now aware of more powerful forces in nature and he finds these disturbing.
10. It marks a sea-change moment in his understanding of the world. The experience changed his views of nature and of the power of imagination for ever. It was an important milestone in his growth as a poet.

My Last Duchess

1. That he is a born performer and likes to show off; he thinks he can read people's thoughts – because that's how powerful he is! That he needs to believe he can control every situation, real or imagined.
2. He wouldn't have known what to say, as he isn't a skilled speaker; it would have been beneath his dignity to speak of such things to her.
3. He tells him to sit and look at the portrait (5); he tells him they will go downstairs together (52–53).
4. Yes, because the entire monologue has been about the Duke and he is utterly self-absorbed and narcissistic.
5. The title suggests that the Duchess in the portrait is only one in a whole series of Duchesses.
6. He keeps strict control (he is possessive) over access to the painting; only he can see it or allow others to.
7. Very few lines are end-stopped, and it is significant when they are (e.g. line 45) whereas the rhyming couplets are very controlled and maintained throughout. It suggests the Duke's struggle for control: despite his efforts to control them, his emotions overcome him, shown by run-on sentences and mid-line breaks.
8. In one sense, the Duchess's portrait is just another of his artifacts, comparable to the statue of Neptune. The statue also symbolises his obsession with status, as it has been made for him by a famous artist and he wants to boast about that. Though it may seem like a throwaway aside at the end, Browning gives it special significance by placing it here, and he uses alliteration on 'n' and 't' and 's' to create a rich and smug-sounding description for the Duke to roll around his tongue.
9. Because he has the complete control over the portrait that he desired to have over the living person and it is a tangible reminder to him of his own power; perhaps Browning is suggesting that art cannot be divorced from morality. The fact that the Duke considers himself an art connoisseur yet appears to have no sense of morality carries an implicit criticism of art as an aesthetic exercise.

Pages 38-45 Revise Questions

Page 39 Quick Test Answers
1. A serviceman who has returned home from Iraq.
2. By the speaker's use of colloquial expressions and conversational markers.
3. Because the soldier cannot forget the incident and the dead man continues to haunt him.

Page 41 Quick Test Answers
1. She seems to feel sad about conflict. There is a sense of regret as she watches her son leave, but she doesn't try to stop him. She accepts the idea of conflict, but doesn't want her child to be part of it. He is 'intoxicated' (22) by the idea of becoming a soldier and she respects this.

2. That it is traumatic. Without their loved ones, those left behind are always 'hoping to hear' and waiting nervously for them to come back, or for news of them.
3. The songbird represents her son. It leaves, like he does, looking for freedom. The dove is a symbol of peace, which may suggest she is hoping for peace. It can also be a symbol of mourning.

Pages 43 Quick Test Answers
1. Developing photographs in his dark room.
2. Belfast, Beirut and Phnom Penh.
3. Though they are moved by the photos, their lives do not change.

Page 45 Quick Test Answers
1. Tissue paper.
2. Paper pages in a copy of the Koran; maps; paper receipts; paper kites.
3. Living (human) tissue.

Remains

1. Two from: 'legs it'; 'letting fly'; 'tosses his guts'; 'carted off'; 'end of story'; 'flush him out', 'near to the knuckle'.
2. It has many meanings: it can be used as a verb meaning to stay or to be left. It can be used as a noun, meaning bits that have been left behind or that still exist, or it can mean a corpse. The meanings are relevant because: the incident resulted in a corpse; the killing and the image of the man being gunned down remains in the serviceman's mind; the looter's body left a blood-shadow on the ground.
3. It is not clear, but he may be speaking to the poet.
4. He wants to make it clear that he did not act, or decide to take action, alone.
5. On patrol he walked over the blood-shadow and perhaps this suggests he was able to live with what he had done while he was on duty.
6. He takes drugs and drinks.
7. Though he knows he was acting under orders and that all three of the soldiers took the same action, the last line of the poem suggests that he does feel responsible for the looter's death.
8. The lines suggest that the man they killed is not dead, but lives on in the serviceman's head.
9. It tells two versions of the same incident – first, the speaker's recount of what happened and then again with the scene that is stuck in his memory. This mirrors the way the speaker replays the same scene over and over in his mind.
10. He uses the present tense to describe something that happened in the past to suggest that the incident is continually present for the speaker. This use of the dramatic, or historic, present tense makes events more vivid for the listener/reader.

Poppies

1. A mother of a son who is a soldier.

2. Since the First World War, poppies have become a symbol of remembrance of the sacrifice made in all wars. Their mention here sets the tone and theme of the poem.
3. The gesture, with its link to childhood, would bring up poignant memories for both of them. She wants to remain in control of her emotions and she puts on a brave face for her son.
4. Her feelings are ambiguous: she hopes to hear his voice again (35) and leans against the memorial 'like a wishbone' (32), but because we are not sure whether the son left recently (and may return) or a long time ago (and won't because he is dead) it is not clear whether she really expects him to come back or not.
5. It may suggest that she wishes her son would come home safely; it also seems to suggest that she is bent double with her head down, in grief – or prayer.
6. She mentions two gestures: rubbing noses, running her fingers through his hair, that are things she used to do when he was a child. She mentions his 'playground voice' in the final line.
7. 'flattened, rolled, turned into felt,/slowly melting' (17–18). It suggests she can't remember what she said, only that she said something, and that her emotions were overpowering at this point.
8. 1) When she makes a deliberate effort to control her facial expression (10). 2) When she resists the impulse to rub noses and ruffle his hair (11–16).
9. Two from: pinned; Sellotape; cat hairs; front door; bedroom; tucks, darts, pleats; ornamental stitch.

War Photographer

1. He is likened to a priest performing a religious ritual. He takes his work seriously and works methodically ('ordered rows' (2)). (He thinks his job is important, even sacred. There is a sense that he is performing a religious ceremony for the dead commemorated in the photographs.)
2. It is a place where the most worrying concern is what the weather is like – no bombs, no gunfire, no violence, none of the horrors of war exist there.
3. It is the photograph of someone he does not know, but whom he saw go through the most personal and private of all things: death.
4. The editor sifts through and discards almost all of the images, selecting only those with the most impact for his readers.
5. It suggests that the photos offer readers a momentary respite from their usual Sunday routines.
6. He is in a plane over England.
7. One interpretation is that the photographer is hardened to his job (and it is just a job) such that he is now emotionally disengaged ('impassively' (23)) from his own country because of its perceived indifference.
8. Yes: he has to choose images for his readers that accurately portray the

conflict without prejudicing their opinion and to choose only a few from so many is difficult.
9. He is more affected by the photographs than he was when taking them; that he finds being back in England more disturbing than being in a war zone.
10. The image of the photograph gradually emerges in the development solution, taking solidity. It is a photo of a dead person, so it is as if he is developing a picture of someone's spirit or ghost.

Tissue

1. Reflective.
2. Freedom; transience; power; fragility; transformation.
3. Two from: handwritten pages in the Koran (history and belief); buildings (cities and construction); landmarks and geographical features (geography); paper receipts (consumerism).
4. Paper allows an architect to create something living, that lets in the light.
5. It perhaps suggests the human form.
6. The speaker values transience over permanence; transience is more genuine, as nothing in life ever really lasts, though we might think it will.
7. All the types of paper mentioned in the previous verses: handwritten pages, maps, receipts – all the records and objects that we use to try to control the world.
8. They suggest freedom, impermanence and transience – reminding us that these are the valuable aspects of life.
9. It gives her final thought, of paper turning into human skin, more emphasis and significance.
10. It could be: paper takes away the power of the things that are recorded on them; or perhaps paper encompasses their power. The poet may be saying that everything we need, and everything we need to know is encompassed by our own selves as human beings or that everything that humans have created is insignificant compared with the greatest of all creations, Man.

The Charge of the Light Brigade

1. Out of a sense of patriotism and to boost national morale during the war.
2. 'Someone had blunder'd' (12).
3. 'Was there a man dismayed?' (10); 'When can their glory fade?' (50); This rhetorical device gives a public tone to the poem – as if the poet is giving a public oration to a sympathetic audience. It helps to involve the reader in the narrative.
4. To focus the reader's attention on the bravery of the cavalrymen and the strength of the opposition they faced in the charge itself. (He does not want the reader to consider the political context).
5. Each line has six beats that are arranged

in sets of three to mimic the sound of horses' hooves in the charge and the sound of a beating heart. It gives a relentless, forward motion to the poem and perhaps suggests the unquestioning and courageous way the cavalry obeyed the order to charge. The rhythm changes in important lines in the poem such as 'All the world wonder'd' (31) and 'Rode the six hundred' (16) (echoed in 'Noble six hundred' (55)). These lines are very short, more reflective moments in the poem.

6. Three from: 'Boldly they rode and well' (23); 'They that had fought so well' (45); 'When can their glory fade?' (50); 'Honour the charge they made!/Honour the Light Brigade, 'Noble six hundred!' (53–55).
7. Suggests the strong organisation and discipline in the ranks of the cavalrymen.
8. The formality gives an official and oratorical tone that is in line with the patriotic theme.
9. 'All the world wondered' (31, 52).

Exposure

1. Physically, the regiment is in an exposed and vulnerable position – they are 'exposed' to icy wintry conditions that are as great an enemy as the German troops – they are emotionally and mentally raw, having been stripped of belief in anything. They are 'exposed' in every sense.
2. Assonance of 'i' in line 1 suggests the sound of the freezing winds as they whistle through; The alliteration of 's' in line 4 creates the sound of nervous whispering as they wonder when the enemy will attack.
3. The metre is unusual, because it has 12–14 beats in each line, which sounds jarring to the ear and evokes the sense of unease the soldiers feel as they wait for the Germans to attack.
4. The repeated line has the effect of underlining Owen's belief that the war is futile. It also stresses the sense of suspended animation of the ongoing waiting situation — the soldiers do not know when the enemy will attack, only that they will.
5. Part II explores the ramifications of Part I. In it, Owen offers his thoughts on the meaning of death for the soldiers in the context of the war, which has stripped them of their beliefs about life.
6. He thinks the soldiers are more likely to die of exposure to the cold than from enemy fire.
7. He calls the fires 'sunk' (26) (dying) and the image is of them as 'crusted [with] dark-red jewels' (27) is suggestive of the bloody wounds of dying soldiers. In one sense the fires of home are tainted by the blood of those who died for them, but the fires themselves are dying because those soldiers no longer believe they will live to see them again. Or perhaps it suggests that those fires are not worth the sacrifice.
8. The rhymes are not 'perfect' rhymes to underline the sense of nervous, off-kilter unease in the soldiers as they wait

for enemy attack.
9. It is a powerful image of the very liquid of the eyes turning to ice because it is so cold. It is the more powerful because it is ambiguously placed. Does it refer to the eyes of the dead or to the eyes of the burial party? If the latter, does it mean that those in the party have become emotionally deadened by the terrible job they do – and their eyes are 'ice' because they no longer feel anything?

Storm on the Island

1. As a powerful and dangerous force.
2. 'as you see' (4) and 'you know what I mean' (7) directly address the listener/reader; It suggests that this is one side of a conversation and adds real-life immediacy, and truth, to the speaker's words.
3. The oxymoron of the phrase juxtaposes two ideas about the sea: that it is very powerful 'exploding', but usually benign 'comfortably'. The idea is revisited in the later image of the sea as a tame cat turning savage.
4. He likens the storm to an air attack by enemy aircraft.
5. Resilient and also brave.
6. He realises that because the storm is composed chiefly of wind (air), its attack is composed of air (nothing).
7. By his use of the pronouns 'we' and 'our'.
8. The word is chosen to link with the military imagery used to describe the storm. The islanders are a community offering a collective resistance to the storm and the speaker tells his listener that though trees and sea might be expected to offer 'company', i.e. military support, in fact they don't offer any 'company' at all.
9. It is positive, but the negatives show that he is very aware of the dangers they face and the fact that they face those dangers alone.
10. Possibly. Some people have said the title is a play on words: 'Stormont Ireland' and that the 'empty air' of the storm refers to empty words and promises of politicians.

Bayonet Charge

1. It plunges the reader straight into the action, very abruptly.
2. The monosyllables in 'field of clods' (3) and 'green hedge' (3) and the assonance of 'o' in 'across', 'of' and 'clods' (3) make us hear how he stumbles in the mud.
3. Dazzled is often used in a positive sense to describe the effects of sunlight, but here it is gunfire that makes the hedge sparkle.
4. Patriotic pride; and fear.
5. 'his foot hung like/Statuary in mid-stride' (14–15)
6. It may suggest that conflict is unnatural and that its violence causes terrible damage to natural beauty.
7. He questions what greater purpose is being fulfilled by his being in that place at that moment.
8. The terrified hare.

9. The 'blue crackling air' (22) is a powerful image of the continuing gunfire that envelops him and echoes 'dazzled with rifle fire' in line 4. Likening his terror to 'touchy dynamite' (23) suggests that the soldier's emotions are dehumanised and echoes the image of his fear as 'molten iron' in line 8.
10. The use of the word 'awoke', the way the poem starts so abruptly with 'Suddenly', together with the simile in lines 12–14 mean that we could interpret the poem as a description of a waking nightmare – or a recurring traumatic memory.

Page 55 Quick Test Answers
1. When she was a child.
2. There has been some kind of conflict and a repressive regime has taken over.
3. She feels under threat: accused of being 'dark' and absent.

Page 58 Quick Test Answers
1. The suicide attack. He was a kamikaze pilot.
2. Because of the disgrace of aborting his mission.
3. It was never spoken of.

Page 61 Quick Test Answers
1. School history teachers and historians who promote a white version of history.
2. It suggests the control that white people have over the history that is taught in schools. It may suggest that Agard felt nagged at by his teachers.
3. That the speaker feels his view of history and his own identity, has been distorted.

The Emigrée

1. It could symbolise anything the speaker holds dear – her family, a lover, religious belief, even identity itself.
2. 'as time rolls its tanks' (10); 'frontiers rise between us' (11).
3. Examples: 'the bright, filled paperweight' (6) emphasises the clarity and vividness of her visual memory and how the image seems frozen in time, unchanging; 'branded by an impression of sunlight' (8) suggests the image she has of the city is of a bright, searing heat, lit by a light so intense that it has burned its image into her own skin, taking ownership of her.
4. That she has no formal nationality and cannot travel anywhere.
5. Because it no longer exists, as it was then. It has changed.
6. Her language and her imagination.
7. Though she is in a dangerous position (no passport and under threat of death) the speaker's tone is one of quiet defiance and certainty.
8. 'They' are clearly the speaker's enemies and they pose a threat: they could be her own countrymen or they could be people in the place she lives now.
9. She perhaps knows things about the city that 'they' want her to tell them, but her instinct is to stay loyal to the

city, to protect it.

10. Probably not. The words 'I am told' (4) and use of 'may' (7, 15) indicate doubt in the speaker's mind. Lines 5–6 also indicate that she resists what she has been told ('The worst news').

Kamikaze

1. He shaved his head, said an incantation and laid his Samurai sword in the cockpit.
2. Fishing boats.
3. That the speaker is speculating about what her father thought, to find reasons for his decision.
4. The beauty and freedom of nature.
5. Perhaps he thought about his own children, waiting for him to return.
6. He was a fisherman.
7. He thought of the beauty and freedom of nature; he remembered his brothers and his father and thought of the family he'd left behind.
8. They lived in silence. His wife never spoke in his presence and neighbours ignored him.
9. The speaker attempts to find reasons for his decision, but refrains from making a moral judgement either way. She simply states what she knows to have been the repercussions of his action. Readers are left to make up their own minds.
10. That her father thought he had a choice to carry out or abort the mission; but in fact his choice was between suicide-with-honour and life-with-dishonour, (which was also a kind of 'death'.)

Checking Out Me History

1. At different times the tone is annoyed, mocking and outraged.
2. The tone changes from anger to reverence; from mocking to serious.
3. By putting them in the same category as the real figures from history, he undermines the authenticity of the latter.
4. It clearly evokes the speaking voice of a black person with a Caribbean English dialect – someone who is proud of their cultural heritage.
5. Repetition is used to express his anger in the quatrains ('dem tell me') and to reinforce his own learning (and his teaching of the reader) about history in the free verse, italicised stanzas.
6. The tone changes from one of oppression at the beginning to a positive tone of independence; 'checking out' and 'carving out' show the speaker taking active charge of his own identity.
7. At the start of the poem we assume it is school history teachers, but it becomes more than that as the poem develops to mean anyone who wishes to impose an exclusively white culture and identity.
8. Two from: Toussaint L'Overture; Nanny de Maroon; Mary Seacole. He uses positive imagery and emphasises their achievements.
9. A sense of their own identity.

Remains

1. To deal with looters who were raiding the bank.
2. The looter's bullet-riddled body.
3. It may mean that he can't remember their names, that their names are not important or he doesn't want to implicate them by naming them. However, he wants to be clear that they were all acting together.
4. It has several meanings and may refer to both the serviceman and the victim of the shooting: the looter's 'remains' (or corpse) are what 'remain' in the serviceman's mind.
5. He uses violent verbs to convey the violence of the shooting ('rips through' (9); 'torn apart' (23)) and the way the incident comes back into his mind ('bursts again' (21)).
6. The shooting is described twice to mimic the way that it has taken hold of the serviceman's mind. He re–experiences the incident, over and over again.
7. It does shift subtly through the poem – first, they are acting under orders ('sent out' (1)); then 'all three' (7) of them make the decision to fire on the looter. But by the end, the speaker has put his own part in the shooting and made it personal ('his bloody life in my bloody hands' (30)).
8. Post-traumatic stress disorder (PTSD).
9. For the speaker – for the trauma he continues to experience as a result of the shooting. The poem is very graphic in its description of the way the looter died, but it also focuses on the speaker's reaction to the violence of his death. This makes it shocking to read, but our sympathies are perhaps guided more towards the living victim than the dead one.

Poppies

1. Brave because she did not show her son how upset she was to say goodbye.
2. One from: 'steeled the softening/ of my face.' (10–11); 'I resisted the impulse' (14).
3. 'released a song bird from its cage' (24).
4. 'bandaged' (7); 'steeled' (10); 'my stomach busy/ making tucks, darts, pleats' (27–28). (Any two.)
5. It indicates her son's excitement about what the future holds for him.
6. She wants to acknowledge to him that she knows he is a responsible adult now / she doesn't want him to be upset or emotional.
7. The references remind us of the domestic setting and associate the mother with the home (This also contrasts with the large and military world that her son is entering.)
8. She goes to her son's bedroom and then visits a war memorial at the top of a hill.
9. In his bedroom she releases her emotions (suggested by the image of the song bird in line 24) and then she describes how her stomach was in knots as she walks to the war memorial.

10. Perhaps a mixture of sadness, loneliness and hope.

War Photographer

1. It suggests that the photographer takes it seriously and it is almost as if he is commemorating the dead he has photographed by developing the images he has taken of them.
2. Each verse is like a snapshot of a scene: the first verse sets the scene in the dark room; the second verse has him getting the solutions ready and gives an insight into his state of mind; in the third verse the images bring back memories; the final verse has the editor choosing the photos and the photographer takes off for another job.
3. He feels it is a responsible job and a necessary one; but at the same time a part of him despises the way the photographs are of only momentary interest to readers and the fact that he earns his living this way.
4. His job is important as it is a record of what is happening yet he still feels he should ask for permission before intruding. However, he has to do this by signs, as he can't speak their language.
5. It indicates that the photographs are starting to appear as they develop; it also shows a change of thought in the photographer as the images bring back memories.
6. It describes the physical state of the photographs as they move under the developing solution; it also may suggest that the photographer is viewing the images through tears.
7. The word suggests he is emotionally hardened to the job he has to do or it may be that he feels emotionally disengaged from his homeland which does not care about the conflicts he records.
8. They are essentially downbeat and bitter; The final line is condemnatory of media presentation of, and public reaction to, the photographic horrors of war.

Tissue

1. Tissue paper.
2. It lets the light through; it can change our view of things.
3. Paper is used to record information: measurements, facts, plans, maps and receipts.
4. It lets the light and sun through – it is thin and fine; it can be blown away; it flies off like a kite.
5. It suggests that creation would be a violent, inspired and sudden act (a 'breakthrough'); it recalls the references to paper and light (and the repeated phrase 'shine through') earlier in the poem.
6. In Christianity, God is sometimes referred to as the 'Great architect of the universe'.
7. The grocery store receipts in the sixth verse; 'fly our lives like paper kites' (24); 'the shapes that pride can make' (30) in the eighth verse suggest criticism.
8. The speaker's attitude is calmly

philosophical and it could be argued she is idealistic, allowing a dream-like state to overtake her. Her use of modal verbs give a sense of uncertainty and possibility helps to suggest this.

9. The poet uses many short phrases piled up on one another like the layers of paper in the seventh verse. Like the tissue paper, and like human skin that is made up of layers, the poet layers phrases and a consideration of four distinct types of paper records in verses 1–6 to culminate in the creation of the material used in the final 'grand design' (32), the human form in the final three verses.

Pages 69-71 Review Questions

The Emigrée

1. Through her memory, language and imagination.
2. Because she no longer lives there and it has changed and now she herself is under threat.
3. It has been taken over by some sort of repressive regime.
4. Because the speaker has no passport and can't get back to it herself.
5. The poet uses repetition of 'sunlight' and 'city' as symbols of the speaker's feelings about her past. Sunlight represents freedom and optimism; her city represents her childhood memories, a source of inner strength and resistance against oppression.
6. The speaker associates sunlight with the city of her memory and her childhood, when she was free. It is repeated to show that she often draws on those memories and sunlight (perhaps) represents a hope that she needs to keep alive.
7. It may refer to the colour of her skin, her attitude or beliefs, believed by 'them' to be 'dark'.
8. It links to the childhood memories that she drew on in at the start of the poem – the city for her is part of her childhood memories. It also emphasises the docility of 'her' city, compared to the conflict and repression that she is told exists there now. It also underlines the loving relationship she has with her city.
9. The speaker is stuck, unable to travel, with no passport, and not free, although the city she is in is 'free (23)'. She is under threat, possibly of death and accusations. She may be facing a trial ('evidence' (25)) and she may be in prison ('city of walls' (21–22)).
10. The city may symbolise a lover, a religious or political belief, childhood or family.

Kamikaze

1. He is a kamikaze pilot getting ready to set out on his final mission.
2. It is possible he saw the shoals making patterns in the sea; The speaker imagines he did, because it explains how he made the connection between nature's beauty and the memory of his own brothers and father.
3. It reminded him how his brothers used to wait for their father to return from fishing.
4. He was also engaged in a dangerous job, but unlike a kamikaze pilot he would return safe.
5. The speaker's father was dishonoured: ignored by his own wife and shunned by his neighbours.
6. The speaker offers a train of thought that her father 'might have' had: the fishing boats, the shoals of fish, the brothers waiting for their father – leading him to think perhaps of his own children waiting for him and causing him to turn around.
7. To emphasise the speed, motion, freedom and beauty of the natural world.
8. It offers a concluding thought from the speaker: that her father's life was so miserable afterwards that he must have wondered whether he might have been better off completing the kamikaze mission.
9. The speaker uses italics to speak from her own memory, recording what she knows to be fact (as opposed to the rest of the poem which is speculation on her part).

Checking Out Me History

1. How there are different versions of history, culture and identity; the importance of finding our own identity and our own history; how control of the past means control of the present.
2. Rhyming quatrains that start 'dem tell me' are contrasted with the free verse of the italicised stanzas.
3. It gives the poem a powerful impetus and drive, mimicking spoken language; lack of punctuation may suggest a lack of due reverence for the rules of 'language' / the speaker is taking control of his own language and voice as he 'carves' his own identity.
4. The strong beat of the repeated monosyllables expresses the speaker's anger and scorn towards the history he has been taught.
5. Hey Diddle Diddle.
6. Seacole was a Jamaican-born woman who provided food and lodging for convalescent officers in the Crimean war; The poet compares her with Florence Nightingale.
7. Metaphor is used in the free verse, italicised verses (6 and 9); He chooses to use it here to present the figures from black history in a positive light.
8. For example, Robin Hood and Dick Whittington. Adds humour; undermines the 'authenticity' of the 'real' historical figures he has been taught about. It suggests that to the speaker, the story of Robin Hood is as relevant to him as the historical facts about Lord Nelson.
9. It is used to show the speaker's own culture and difference from the history that white culture taught to him.
10. The final two lines. The verbs 'checking out' and 'carving' show he is actively taking control of his own heritage and identity.

1. Ozymandias/My Last Duchess
Suggested content:

- Both poems explore the effects of power: Shelley's poem leads us to understand that even absolute power is fleeting and transitory; Browning's poem explores its ruthlessness. The subjects of both poems – Ozymandias and the Duke – share characteristics of arrogance, pride and egotism associated with the exercise of absolute rule.
- Consider differences in narrative framing: Browning uses the persona of the Duke himself as speaker so his brutality and misuse of power is gradually revealed over the course of his monologue. Shelley uses narrative framing to create distance – his is a story told to the narrator by a traveller and we hear Ozymandias's own words third-hand, through the inscription on his pedestal. We pass judgement on Ozymandias's cruelty and tyranny through the sculptor's (and poet's) artistry.
- Both poets choose to set their poems in an era very distant from their own: Browning's in the Renaissance; Shelley in an 'antique land'. Both poets were critical of power, but felt it was safer to critique it at a distance, though the political resonance of their ideas was very current.
- Compare structure and language: Shelley's critique of power is a sonnet, a form traditionally associated with love not political ideas, but one which shows the poet's skill; Browning chooses the dramatic monologue so that the Duke condemns himself through his own words. Everything he says about the Duchess shows his obsession with control and his narcissism. Like Ozymandias he considers himself an absolute ruler whose words and actions inspire fear and dread. Browning cleverly uses the tightly controlled rhyming couplets to show the Duke's obsessive character fighting against his passions, indicated by the frequent use of enjambment.
- Consider the use and effect of imagery: the statue in Shelley's poem is a symbol of the transience of power; in Browning's poem, the portrait of the Duchess symbolises its ruthlessness. Both the Duke and Ozymandias have artefacts created that reflect, in their different ways, what they view as their own unchallenged power. The irony is that each of these artefacts in fact undermines those views – for in each case, the real power lies with the artists that created them.

2. London/The Emigrée
Suggested content:
- Compare the voice and perspective of the speakers: Blake writes in protest at the social injustice he sees all around him in London. He writes as a social commentator, with anger and sorrow and with the desire to wake people up to the effects of oppression. Rumens writes about political oppression through the voice of a speaker who has been forced to flee from her home country. She is powerless to do anything to change it, but puts up personal and inner resistance to its effects. Both write in the first-person, which gives immediacy and directness to their messages.
- Both speakers write about oppression, but in different ways. Blake sees pain and misery everywhere he looks and launches an attack on those things he sees as responsible – the church and the state. Rumens' speaker offers passive resistance, refusing to believe the stories she is told about her home town and retaining her own sense of identity through language and memory. Blake sees victims of oppression and protests with a call to action, but Rumens only defends her own rights to be herself.
- Compare the use and effects of structure and rhythm: Blake uses unrelenting rhythm, regular metre and repetition to drive his message home. The reader feels his passion in their force. In 'The Emigrée', the female speaker is gentler in her use of rhythm that is similar to the cadence of ordinary language. The effect of this is to give a sense of quiet power, truthfulness and regret. The structure of the two poems is completely different, but each is a powerful response to oppression.
- Consider the use and effect of language and imagery: Rumens' repeated image of sunlight (the final word of every poem) becomes symbolic of her belief in the city of her memory and that is a symbol of her own identity. Her image of the city as a doll shows the power she has to retain her own identity and how she protects it. It is her talisman against the oppression she currently faces. Blake uses violent imagery to shock and involve the reader's emotions about the effects of oppression (blood, curse, blasts, blight and manacles).
- Consider the differing contexts of the poems: Rumens is not writing about a particular conflict, but uses her speaker to comment generally about the effect of conflict on individuals. Blake was writing about the effects of the Industrial Revolution on society. Blake is speaking for, and to, society in general, while Rumens considers the effect of oppression on an individual.

3. The Emigrée/Poppies
Suggested content:
- Consider the speakers' attitudes to loss: Both speakers are female and speak in the first-person, giving insight into their thoughts about loss. Rumens' speaker has lost the country of her birth and her identity is under threat, but she clings to her childhood memories and her language to preserve her identity.
- Contrast the voice and perspective: the city of the emigrée's memory has gone and her tone is regretful, but she does not grieve because she keeps the memory of it alive and draws strength from it. She calls the city docile because she has made it her own, taken control of it. The mother in 'Poppies' feels terribly the loss of her son and laments him. She remembers certain details of their last meeting – his hair, the cat hairs on his blazer – things that stick in her memory because of her heightened emotional state.
- Presentation of theme of loss: Both speakers mourn the loss of something dear to them and both powerfully cling to the memory and draw strength from the memories: the emigrée uses her imagination to enable her city to visit her like a doll or a pet or a lover, giving her inner freedom, though she is under threat and has no way of leaving the new city she is in. It is ambiguous what the 'city' symbolises. It may be anything that someone holds dear and which people try to take away – a failed love affair, for example, or religious belief. In Weir's poem the loss is a mother's loss of her son, but it also is ambiguous. We are not sure whether the son is dead or not. The poets offer differing interpretations of the effects of grief and loss on individuals.
- Consider the use and effect of language: both poets offer a female perspective on grief and this is reflected in the feminine imagery chosen by their speakers. Rumens: 'like a hollow doll', 'comb its hair and love its shining eyes', 'takes me dancing'; Weir domestic imagery of sewing ('tucks, darts, pleats') and motherly actions of running fingers through hair, smoothing a collar and picking off cat hairs with sellotape. Both poets use imagery to suggest the intense emotional state of the speakers and how imagination is heightened by grief.

4. Tissue/Checking Out Me History
Suggested content:
- Themes and attitudes: both speakers understand the power of language to transform identity. Agard reclaims his self-identity through learning about the history of his own culture and the deeds of famous black people like Mary Seacole and Toussaint. He is angry that he was not taught about them. Dharker's speaker sees language as slippery, but strong. It controls and creates us. Both poets view the control of self-identity in relationship to having power over the cultural and historical contexts that shape us.
- Speaker's voices and perspectives: Dharker's poem is a kind of soliloquy of inner thoughts that present a series of images to persuade the reader about the impermanence of concrete things: the landscape, buildings, things we buy. The voice is tentative, using the subjunctive and modal verbs (If ... could/ might) as if sensitive to its own impermanence. Agard's speaker by contrast is strident and angry, protesting about the way his culture was omitted by the education he received and showing how he is taking that power back. Both poets understand the importance of the written word as a force for oppression or freedom.
- Language and imagery: Dharker uses a metaphor of paper and its power to change the way we view the world. The

paper may represent displacement from own identity and how if individuals choose to let go of things that control us (our history, nationhood, consumerism/ capitalism) we might be free to create our own self-identity ('raise a structure never meant to last'), and even to accept our own mortality. Agard mixes storybook characters with real historical figures in a humorous way, but with serious intent. He uses startling and colourful metaphors for the famous black figures he admires ('Fire-woman', 'See-far woman' for Nanny de maroon, for example).

- Both poets challenge conventional ideas about history and nation in urging us to find our own self-identity and this is reflected in the structures they choose. Dharker's poem is organised in conventional four line verses, but her sentences, like her thoughts, flow across their boundaries and few lines are end-stopped. She finishes with one line on its own for emphasis, that changes the meaning of the whole poem. Agard uses two types of verse (rhyming quatrains and italicised free verse) and presentation to show the difference between the culture he was taught and that which is really his. His repeated 'dem' and 'dem tell me' reveals his anger at being prevented from learning about his own culture.
- Both poets talk about the importance of claiming self-identity; both champion the individuality and freedom to find it.

5. Bayonet Charge/Exposure
Suggested content:
- Consider the use of voice and perspective: Owen writes personally and is the speaker of the poem. This gives the impact of immediacy and raw emotion, as if the reader is privy to his inner thoughts and emotions. Hughes uses a third person narrator who goes inside the soldier's head. In a sense, Hughes writes for every soldier as his soldier is unnamed.
- Compare themes and setting: Both set at the front in World War 1. Owen describes the tension of waiting for battle to start; Hughes's soldier is in the midst of a charge. Both convey the terror of war, but in different ways. For Owen, the horrors of war are daily in front of him in the icy corpses cleared away every night. The repeated phrase 'nothing happens' underlines the futility of war. Hughes's soldier also has a moment of contemplation, even under fire – he wonders what he is doing there, and how he got to that place in that moment. For Owen, the suspended time between engaging with the enemy makes him realise how pointless it is even to go home, for everything has changed – war has changed him, and everything, for ever.
- Compare the use and effects of structure and language: Hughes' imagery is clumsy compared with Owen's, which weaves images into the fabric of the poem. Hughes uses natural images to contrast with that of warfare. His 'yellow hare' is an image of war's

destructive impact on nature. Owen's language is more delicately and hauntingly beautiful and poignant with the way it conveys the sense of cold tension in every line. He employs half-rhymes to suggest the tension and unease of the waiting soldiers.
- Contrast the contexts – Hughes's poem was written from a distance of 40 years from the end of World War 1 whilst Owen's was written at the Front, yet it is the later poem that strives to give a sense of the battle's violence, whilst Owen's is the more distanced and philosophical in tone. If you are really in the heat of battle, you are more inclined to seek to escape it in your thoughts. Both poems reflect this, but have different ways of conveying it.

6. The Charge of the Light Brigade/Bayonet Charge
Suggested content:
- Consider the context: Tennyson was a Victorian Poet Laureate and writing in an official capacity. Therefore he presents the patriotic version of a specific battle, that in reality was a staggering defeat. Hughes is not writing about a specific battle, but more symbolically. His view of patriotism is perhaps coloured by the fact that his father was a World War 1 veteran and in the poem he is trying to get inside the head of a soldier at the front.
- Both poets use the third person, but Tennyson's gives no insight into individual thoughts or feelings of the cavalry. Tennyson admires their courage in the face of impossible odds and he says their patriotism is to be admired and honoured because they unthinkingly followed orders with no thought of the danger. Compare with Hughes's soldier who stops mid-charge and questions what he doing there. For this soldier, patriotism ('King, honour, human dignity, etcetera') is meaningless and irrelevant in the panic of battle.
- Compare the use and effects of rhythm and structure: Tennyson's 2-beat rhythm drives the poem along, suggestive of hoof-beats and emphasising the excitement and glory of battle. By contrast Hughes's free verse, enjambment and use of violent verbs evokes the soldier's panic-driven and terrified charge. He is out of control, compared to the orderliness of Tennyson's cavalry.
- Consider the use and effect of imagery: Tennyson uses grand religious-sounding names and personification to make the charge sound glorious and the valley a place of momentous historical importance ('jaws of death', 'valley of death'). Contrast the natural imagery of Hughes, who wants to show the terrible cruelty of the war's reality that undermines patriotism: the yellow hare, the furrows and hedges of the countryside contrast with the terror of rifle fire.
- For Hughes's soldier, patriotism is a 'luxury' that the terror of battle cannot afford. For Tennyson, blind patriotism (and the courage it requires) is to be honoured.

7. Exposure/War Photographer
Suggested content:
- Consider voice and perspective: Owen is tired and resigned to the futility of the battle he is waiting to re-start. In the trenches, on the frontline, the futility of the whole conflict is painfully clear to him. His repeated refrain 'nothing happens' suggests that the war itself will achieve nothing and therefore the current conflict is pointless. Duffy's photographer is also cynical about war, but his feelings are mainly directed towards the readers of the media in his own country: 'they do not care' about the conflicts that he records with his camera.
- Consider the use and effect of imagery: both poets use starkly visual images to evoke the futility of war. In Duffy's poem, the developing photo 'twists' into a ghostly shape and reminds the speaker of the pain and grief of innocent people caught up in conflict. The line 'running children in a nightmare heat' brings to mind a famous Vietnam war photograph. He also compares the gentle climate of England with the scorching heat of the battlefield, reminding us that far-off battles have little impact to Sunday supplement readers.
- The title of Owen's poem, 'Exposure', has several meanings, one of which is the process of creating photographic images and the pictures he creates of his view from the trenches: flares, driving snowflakes and frozen corpses are powerfully visual. His image of the fires of home, 'crusted dark-red jewels' are not welcoming, but are reminders of bloody wounds on soldier's corpses.
- Contrast the mood and tone: Owen's nihilism reaches a peak in the seventh verse, when he says that everything, even his belief in God, has been destroyed by his experience of war. War has stripped him of everything he thought he believed in and loved: only mice and crickets remain at home, so there is absolutely no point in returning. For the photographer, the way his photos are selected by an editor and then carelessly consumed by readers and instantly forgotten, underlines for him the futility of his own job. Though it is 'necessary', he has to train himself to be unemotional about the pain he sees and 'impassive' about the lack of impact his photographs really have.

8. War Photographer/Remains
Suggested content:
- Consider the themes: Armitage writes about a soldier with PTSD who is haunted and permanently damaged by his involvement in a shooting of a looter. Duffy's speaker is disgusted by the voyeuristic elements of his job as a war photographer – observing the brutal effects of conflict without being involved – and of those who look at his photographs without really caring about the conflicts they record.
- Tone and mood: Both speakers are realistic about the effects of war and feel powerless to change their

situations. The soldier in 'Remains' is trapped in the events leading up to the shooting and continually replays them in his mind, however hard he tries to forget. Armitage powerfully evokes the horrific effects of war on individuals. For Duffy's photographer, forced indifference is his only defence against the futility he really feels about his job and the public he takes the photographs for. Though his job is necessary ('to do what someone must') his photographs don't change anything.

- Compare the use and effects of language: Armitage's soldier speaks very colloquially ('so all three', 'he's there on the ground') and uses the present tense to show that this scene is permanently present for him 'here and now'. This gives readers the sense that this testimony is real and so his un-poetic, plain descriptions of the body 'sort of inside out' and 'the image of agony' are very powerful. The photographer uses language to paint visual pictures: 'blood stained into foreign dust' and 'a half-formed ghost' reminds him of the cries of victims. Both use the word 'agony' to convey the horrific effects of war.
- Contrast the voice and perspective: Armitage uses first-person to give immediacy and impact to the soldier's experience and shows the impact on him by having the shooting repeated - once as a memory and once as a recount of how the memory has become a waking nightmare. Duffy uses the third person to describe the photographer's reaction to his photographs – they also spark memories of pain and suffering, but they are at second-hand, not directly involving the photographer so the reader is placed in a similar position to the speaker – viewing the images without being directly affected.
- Consider the contexts: both poets are writing about recent conflicts (Armitage, Iraq; Duffy lists several 'Belfast. Beirut. Phnom Penh.'). Armitage focuses on the effect of conflict on a single soldier, while Duffy uses the photographer to comment on how we are inured to its horror unless it affects us personally.

9. Kamikaze/Poppies
Suggested content:
- Consider the voices and contexts: In 'Poppies' the speaker is a grieving mother lamenting the loss of her son who has joined the army: Garland's speaker is also an older woman, but she is telling her grandchildren about her father's experiences in war. Both speak with a tone of sadness and regret.
- Both speakers were hurt by war and their male relatives' involvement in it. For Garland's speaker her father's actions in aborting his mission had a huge impact on family life and his reputation in society was ruined for ever. His marriage too was destroyed. It was 'as though he no longer existed'. In Weir's poem it is ambiguous as to whether her son is one of the dead in the 'war graves', but the implication is that just by joining the army, her son is lost to her and she laments his loss. Both speakers

empathise with the male relatives' feelings: Weir's speaker thinks her son is 'intoxicated' with excitement, while Garland's has to imagine what her father thought, as he never spoke of it. She thinks he must have thought it would have been better to die in the plane.
- Compare the structure and language: Both speakers use long sentences, enjambment and connectives to suggest the emotional outpouring of their first person narratives. Weir's is a dramatic monologue, addressing her son as 'you' and continuing to speak as if he is there after he has gone. Garland uses direct speech in italics to indicate the parts that are the speaker's personal experience. The parts of the narrative that she speculates about are reported in the third person and they feel less emotional.
- Contrast the use and effect of imagery: Weir's poem starts ominously with mention of war graves and it ends at a war memorial; there are images of pain ('bandaged') set against hope ('wishbone') and peace ('a single dove'). This use of imagery mirrors the conflicting emotions of the mother who bravely lets her son go without showing him the pain it is causing her. Garland uses imagery to involve the reader in the reasons for the father's decision to turn home: he sees the silver whitebait and shoals of fish and it reminds him poignantly of his childhood and his own father.
- Contrast the impact of the endings: Garland focuses on the terrible impact on her father's life, effectively over after the mission; Weir ends with a mother's lament that the child she knew can never return except in her memory.

10. Poppies/War Photographer
Suggested content:
- Compare the feelings of the speakers: Weir's laments the loss of her son, recalling in vivid detail the day they parted and how she bravely struggled to control her emotions for her son's sake. For her, and all mothers, war means grief and the immense pain of loss. Duffy's photographer also has conflicted feelings – on the one hand he makes his living from images of conflict, but he cannot be dispassionate about the pain and grief he sees.
- Both speakers express pain as an effect of war, but for the mother in 'Poppies' the memorials and graves are symbols of sadness. There is no sense of the glory of war for her, though she recognises that her son feels 'intoxicated' in anticipation of the experience. She can only hope he will return, though her son the child is lost to her for ever. The photographer has first-hand experience of the brutal effects of war, 'blood stained into foreign dust' and though his job is necessary there is a sense that he despises having to do it. The images he develops are a graphic record of the pain of war: 'he remembers the cries/of this man's wife'.
- Both poets explore the agonies of war, for those who experience it first-hand and those who are left behind. Weir's

speaker does not judge the rights or wrongs of conflict, but only records its effects on mothers. Duffy's poem bundles all wars together 'Belfast. Beirut. Phnom Penh. All flesh is grass' as if the reasons or contexts for conflict are irrelevant – they all result in death and misery.
- Consider the poet's methods: Duffy uses the third person to get into the mind of the photographer and explore the unusual position he has of recording, but not participating in conflict. By doing this she can deliver a judgement on Western media's consumption of war photography without really engaging with it. In contrast, Weir's use of the first person and dramatic monologue forces the reader to empathise and engage with the mother's suffering.

11. Remains/Bayonet Charge
Suggested content:
- Consider the themes and attitudes: both poems explore the effect of conflict on the mental state and trauma experienced by individual soldiers. Neither speaker is named, and anonymity's important: the soldier in 'Bayonet Charge' could be said to represent every soldier under fire. Armitage draws attention to anonymity when the speaker refers to 'myself and somebody else and somebody else'. The names are not important: this could happen to anybody. In the second verse of 'Bayonet Charge' the soldier pauses for an instant to consider what he is doing there, but then carries on his charge, underlining the idea that the battlefield is not a place for reflection. Armitage's soldier opens fire on the looter without knowing whether he is armed or not and the poem records his later struggle with PTSD.
- Contrast the speakers' voices and perspectives: Armitage uses a common soldier's casual idiolect in phrases such as ('one of my mates' 'carted off' and 'letting fly') which makes it sound like he is chatting normally though the event he describes is horrifying. The contrast makes the shooting and its effect on the soldier more shocking to the reader. Hughes's use of the third-person to narrate in his poem has less impact since there is distance between the soldier's actions and thoughts and the reader, though the speaker is omniscient, knowing the soldier's thoughts, his terror and his panic. This puts the reader in the position of observer. By contrast, Armitage's use of dramatic monologue with an assumed listener/interviewer draws the reader closer.
- Compare use of language and imagery: Hughes contrasts the effect of violence on nature with war and how it dehumanises the soldier. The image of the yellow hare leaves a lasting impression on him: though written in the present tense, it is not clear whether the narrative is a memory or a waking nightmare that recurs for the soldier after the event – of the same kind that Armitage describes. Armitage's use of the soldier's own language and syntax to describe the shooting makes it more real and shocking for the reader ('sort of

inside out', 'the image of agony', 'tosses his guts', 'So we've hit').

- Compare structure and form: Armitage cleverly repeats the same incident twice in the poem, so that its structure mirrors the way the soldier helplessly replays the incident himself, over and over. Hughes uses enjambment to add to the sense of forward movement of the soldier, as if like a clockwork toy he is unable to stop in the charge.
- Contexts: Hughes wrote the poem forty years after World War 1 where it is set and there is imaginative distance in his use of the third person to describe the soldier's actions, feelings and thoughts. Armitage interviewed veterans from the Iraq war so his poem is based on real testimony. The poet captures the phrasing and word choice of the speaker.

12. Storm on an Island/Ozymandias
Suggested content:

- Themes and attitudes: the traveller's story in 'Ozymandias' is set in a desert, whereas Heaney's poem is set on a remote island. Both are isolated places that show the power of nature as invincible compared to human power, which is fragile. Shelley is saying man's power is transitory and time conquers it. In 'Storm', the islanders' only defence is to hunker down and wait ('sit tight') until the violent storm has passed.
- Speakers' voices and perspectives: Heaney's speaker is an islander who uses personification to describe the storm and an extended metaphor of an enemy bombing raid to convey its violence. He speaks conversationally, but with experience, and has great respect for nature's power. Shelley uses the character of an incredibly powerful and cruel ruler who boasted of his invincibility. Ironically all that is left of him now are the remnants of a statue. The desert, and time, have vanquished him.
- Language and imagery: Heaney's speaker shows his fear of and vulnerability against nature's power in phrases such as 'full/blast' and 'The thing you fear'. He knows they are entirely at the storm's mercy though he attempts to control the storm by language, likening it to a 'tame cat/ Turned savage' and tries to appear brave and resilient to his listener with conversational interruptions such as 'you know what I mean'. Shelley uses juxtaposition 'colossal wreck' and 'Nothing beside remains' to ironically underline the fact that Ozymandias's power, pride and arrogance are insignificant now. His name also provides a clue with its literal meaning 'ruler of nothing' which contrasts with his self-proclaimed title, 'king of kings'.
- Structure and form: Shelley's poem is a formal sonnet, traditionally associated with love poetry, but here used to convey a political message about the nature of human power. Both poets use iambic pentameter which gives a solemn and weighty, even grand, tone to their words.
- There is pride and a certain grandeur

in the way the islander describes how his community withstands nature's immense power. At the same time he is awed by its violence. Heaney manages to convey several emotions in his presentation of power. Ozymandias's power on the other hand produced arrogance and pride in the ruler. His intention was to inspire awe in others, but Shelley's presentation completely undermines this.

13. My Last Duchess/Checking Out Me History
Suggested content:

- Consider the speakers' feelings: Browning's speaker, the Duke, is boastful and arrogant, sure of his own power, cultural identity and sense of entitlement. He boasts of his lineage, whereas Agard's speaks in his own voice and is angry that his education denied him access to his own culture and history. He intends to find it for himself. The very fact that he is writing the poem is an indication that he is finding his own voice. Note also the absence of the Duchess' voice in the poem.
- Compare the use and effects of structure and form: Browning's poem is a dramatic monologue and the Duke controls it. He rewrites history as he speaks to the envoy, talking about his duchess in the painting, as if he owns her history as he now owns her image. He speaks in tightly controlled rhyming couplets, indicating his desire for rigid control, but these are subverted by enjambment that break up the rhythm, showing how he struggles to control his feelings. By contrast, Agard uses two different structures: rhyming quatrains for the white history he was taught and longer, freer and italicised verses for the famous figures from black history. He associates white cultural power with traditional poetic forms, but he rejects these – and indeed punctuation altogether – when it comes to expressing his own black identity.
- Consider the use and effect of language – the Duke speaks informally, almost conversationally self-interrupting in an attempt to make the envoy feel at ease. He speaks in different voices and tones in what is essentially a performance to show his own power and impress the envoy with his possessions and with anecdotes that illustrate his absolute control. Agard uses a Creole idiolect to emphasise the difference between his identity and that of the white culture which has been imposed on him. His language is also his power, unique to his culture and his identity.
- Contrast the voice and perspective: Both speakers equate culture, heritage and history with power, though they come from completely different centuries. Browning chose to explore the theme of power and tyranny by setting his poem in Renaissance Italy, a time far distant from his own Victorian age. Agard recalls different historical eras in a quest to reclaim his real culture. Though Agard's tone is often

humorous, his message is a serious one. In a similar way, though the Duke pretends to be jocular, his intent is deadly serious.

14. Checking Out Me History/Extract from The Prelude
Suggested content:

- Both writing about their own history: Agard protests about the version of history he was taught, which is not his own black history, and how this resulted in a need for him to find his own culture and identity. Wordsworth's is a childhood memory and recalls an important moment of insight that shaped his growing identity as a poet. Both in a sense rewrite their own history as older poets with the benefit of hindsight.
- Voice and perspective: Agard is angry about being prevented from learning about the history of famous black people and sees this as evidence of white cultural oppression. He alternates using a sarcastic, humorous tone for the white historical figures with a more serious, respectful tone when he talks about the famous black figures from history. Wordsworth starts with a tone of nostalgia and calm when the boy steals the boat, but this changes to panic when (in his imagination) he is menaced by the peak. The end of his poem is reflective as he tries to explain the significance of what happened in developing his poetic sensibility.
- Structure, rhyme and rhythm: Agard's passion is clear from the repetitive drumlike beat of the poem and the repetition of 'dem' like the blow of a hammer also powerfully expresses his feelings. He is taking back power over his own identity that he feels was taken from him by the history he was taught. He alternates two types of verse to make clear the difference between the identity he was taught and the one he now claims. Wordsworth's blank verse, use of conjunctions to link events and use of enjambment drives the reader through the narrative, as if travelling with the boy in the boat.
- Language and imagery: Agard uses 'bandage' and 'blind' to suggest that his view of history and of his own identity has been distorted. By the end of the poem he declares he is now 'checking out' and 'carving' his identity: taking action, taking control. His language is unpunctuated and non-standard English, in defiance at the white education he received. Personification gives dramatic impact to Wordsworth's description of the peak that 'like a living thing/ Strode after me'. We need to feel the boy's terror in order to make sense of the effect it had on Wordsworth's adult identity — as a Romantic poet he never forgot this demonstration of nature's power and awesomeness. Nature was no longer friendly and familiar, but was mysterious and immense.

15. Extract from The Prelude/Storm on the Island

Suggested content:

- Both poems portray nature as threatening: Heaney's storm; Wordsworth's peak. Wordsworth's nature initially is benign. He describes the lake and the stars as calm and beautiful, but after the boy's imagination is menaced by the towering peak, nature's 'huge and mighty forms' troubled him. Heaney's speaker views storms as the islanders' enemy. They are so powerful and ferocious that they threaten the islanders' lives.

- Voice and perspective: Heaney's poem is a dramatic monologue spoken by one of the islanders who explains their preparations for storms. The speaker is at pains to stress the violence and power of the natural forces they need to withstand and how they stand alone without any natural defence from trees, stooks or even the sea. Phrases such as 'you know what I mean' and 'as you see' remind us of the situation. Wordsworth, on the other hand, sets down a significant moment in his development as a poet as part of his autobiographical poem 'The Prelude'. It is his attempt to describe and explain one of his 'moments of being'.

- Language and effects: both poets use personification to describe nature's power. In Heaney, the spray 'spits like a tame cat/ Turned savage'. He uses violent verbs and an extended metaphor of an enemy bombing raid like the Blitz to suggest nature's power: 'strafes', 'explodes', 'dives', 'bombarded'. Wordsworth's peak turns into a huge monster: 'as if with voluntary power instinct/Strode after me'. This compares to the lyrical beauty of his earlier description of the way his oars disturbed the night-time lake, 'Small circles glittering idly in the moon,'. Both poets write in blank verse which gives a sense of seriousness and formality to their descriptions. Heaney's assonance (it, hits, spits, sit) and alliteration ('blows full blast') evoke the violence of the storm's wind and rain. Wordsworth's repetition 'huge peak, black and huge' and alliteration 'growing still in stature the grim shape' emphasises the terror and awe that the peak inspired in the boy.

- Compare the contexts: Wordsworth a Romantic poet, is inspired by and awed by nature's mystery and power. Heaney writing in 20th century is aware of nature's violent dangers from his rural farming childhood. Both are sensitive to the rich beauty and raw power of nature.

Glossary

A

Accumulation   listing of ideas which builds up to create a particular effect

Alliteration   repetition of a sound at the beginning of words

Ambiguity   having more than one meaning

Analysis   a close examination of the meaning, style, form and effects of a poem

Annotate   to add critical or explanatory notes to a poem

Assonance   repetition of vowel sounds

C

Caesura   a pause in a line of poetry. Usually in the middle of a line, but sometimes at the start

Colloquial language /colloquialism   informal language; the sort of language used in conversation; may include dialect words or phrases

Conjunctions   a word that joins phrases/ideas and shows the relationship between them (also called connectives)

Connectives   words or phrases that link, develop or contrast ideas; examples are *and, because, such as* and *of*

Contrast/contrasts   the difference (or differences) between two or more things

Culture   the traditions and beliefs of a society

D

Direct speech   the exact words someone has said

Dramatic monologue   a poem where two voices are heard, often revealing aspects of their characters and details of events leading up to the current situation

E

Emotive   appealing to the emotions/evoking strong feelings

End-stopped   the end of a sentence or clause coincides with the end of a line of poetry

Enjambment   where a clause or sentence runs from one line of poetry to another

Extended metaphor   the continued use of a metaphor for a length of time, sometimes over the entire length of a poem

F

First person   '*I*', '*me*' (singular), '*we*', '*us*' (plural).

Form   the physical structure of a poem, composed of line length, rhyme, rhythm and repetition

Free verse   poetry that doesn't conform to any particular form

H

Half-rhymes   an 'imperfect' or near rhyme

I

Iambic pentameter   a line that contains five iambs (an iamb contains one unstressed and one stressed syllable). Close in rhythm to natural speech

Imagery   a collection of devices (including metaphor, simile personification, synecdoche and onomatopoeia) which use language to create vivid visual descriptions

J

Juxtaposition   the placing of (often contrasting) words (nouns and verbs) or phrases close to each other

M

Metaphor   a form of imagery where one thing is said to be another, suggesting similarities between the two

Metre   the pattern of syllables and stresses that create a poem's rhythm, e.g. Iambic pentameter

Modal verbs   used to express degrees of possibility, certainty, intention or necessity; examples are *could, should, might* and *would*

Mood   the feeling created in a poem through word choice and theme

N

Non-standard English   any dialect of English other than Standard English

O

Oxymoron   the use of words which have contrasting meanings in order to emphasise one meaning, e.g. 'A deafening silence'

P

Parallel   two or more similar ideas, phrases or clauses (parallelism)

Personification   a form of imagery that gives animals, ideas or inanimate objects human qualities

Possessive pronoun   a word that shows ownership: '*my*', '*mine*', '*our*', '*their*', '*his*', '*hers*'

Present tense   a verb tense that says what is happening now

Presentation   the style used by the poet to present a poem

Punctuation   the use of marks to separate words into sentences, clauses and phrases in order to clarify meaning

Q

Quatrain   a group of four lines

R

Repetition   the repeating of the same words or phrases to create an effect

Rhetorical   language used for its persuasive effect on its audience.

Rhyming couplet   two consecutive rhyming lines

Rhythm   a pattern of stressed and unstressed syllables that creates a beat

S

Semantic field   a group of words which have a common theme or meaning

Simile   a direct comparison of one thing to another, using the words '*as*', '*like*' or '*than*'

Soliloquy   a speech to the audience, expressing the speaker's thoughts or feelings

Sonnet   a form of poem, often a love poem, consisting of fourteen lines

T

Theme   subject matter; what a text or poem is about overall, rather than what happens in it

Third person   a narrative viewpoint where the narrator is uninvolved and people are referred to as '*he*', '*she*', '*they*', etc.

V

Verbs   words that describe an action or state

Voice   the style and tone (through word choice, rhyme and stylistic features) by which a poet creates the character of a speaker

Index